I0006294

WHAT IS FREE TRADE?

What Is Free Trade?

And Why Does It Matter?

M. Frederic Bastiat

Stonewell Press

ISBN 978-1-62730-093-3

The views expressed in this book are the responsibility of the author and do not necessarily represent the position of Stonewell Press. The reader alone is responsible for the use of any ideas or information provided by this book.

Published by Stonewell Press, Salt Lake City, Utah.

Any additions to original text © 2013 by Stonewell Press. All rights reserved. Printed in the United States of America.

Stonewell Press™ and the Stonewell Press logo are trademarks of Stonewell Press.

www.stonewellpress.com
editor@stonewellpress.com

Publisher's Preface

Frederic Bastiat (1801–1850) is viewed by some as one of the foremost political economists of all time. A native of France, his works were originally written in the French language. Even though his name has been largely forgotten in many circles, his philosophies continue to have a great influence on libertarians, free-market advocates, classical liberals, fiscal conservatives, and those who argue in favor of a limited government.

He wrote for the common man, earnestly seeking to be understood while also desiring to entertain to maintain reader interest. His discussions of money, capital, government, free trade, and the law are so concise and powerful that they have sometimes thoroughly persuaded those who initially held a completely opposite viewpoint. His philosophies for the most part are compatible with John Locke and Adam Smith. His philosophical heirs have included F. A. Hayek, Ludwig von Mises, and Milton Friedman.

Bastiat's works include *Essays on Political Economy, What Is Free Trade?, Economic Harmonies,* and the ever-popular *The Law.*

Even though some of the arguments of this book are presented in Bastiat's historical context, the principles discussed are as timely as the day they were written.

Bastiat tragically died of tuberculosis at the premature age of forty-nine.

CONTENTS

Contents

Chapter 1

PLENTY AND SCARCITY

Which is better for man and for society—abundance or scarcity?

What! Can such a question be asked? Has it ever been pretended, is it possible to maintain, that scarcity is better than plenty?

Yes: not only has it been maintained, but it is still maintained. Congress says so; many of the newspapers (now happily diminishing in number) say so; a large portion of the public say so; indeed, the *city theory* is by far the more popular one of the two.

Has not Congress passed laws which prohibit the importation of foreign productions by the maintenance of excessive duties? Does not the *Tribune* maintain that it is advantageous to limit the supply of iron manufactures and cotton fabrics, by restraining any one from bringing them to market, but the manufacturers in New England and Pennsylvania? Do we not hear it complained every day: Our importations are too large; We are buying too much from abroad? Is there not an Association of Ladies, who, though they have not kept their promise, still, promised each other not to wear any clothing which was manufactured in other countries?

Now tariffs can only raise prices by diminishing the quantity of goods offered for sale. Therefore, statesmen, editors, and the public generally, believe that scarcity is better than abundance.

But why is this; why should men be so blind as to maintain that scarcity is better than plenty?

Because they look at *price,* but forget *quantity.*

But let us see.

A man becomes rich in proportion to the remunerative nature of his labor; that is to say, *in proportion as he sells his produce at a high price.* The price of his produce is high in proportion to its scarcity. It is plain, then, that, so far as regards him at least, scarcity enriches him. Applying, in turn, this manner of reasoning to each class of laborers individually, the *scarcity theory* is deduced from it. To put this theory into practice, and in order to favor each class of labor, an artificial scarcity is produced in every kind of produce by prohibitory tariffs, by restrictive laws, by monopolies, and by other analogous measures.

In the same manner it is observed that when an article is abundant, it brings a small price. The gains of the producer are, of course, less. If this is the case with all produce, all producers are then poor. Abundance, then, ruins society; and as any strong conviction will always seek to force itself into practice, we see the laws of the country struggling to prevent abundance.

Now, what is the defect in this argument? Something tells us that it must be wrong; but *where* is it wrong? Is it false? No. And yet it is wrong? Yes. But how? *It is incomplete.*

Man produces in order to consume. He is at once producer and consumer. The argument given above, considers him only under the first point of view. Let us look at him in the second character, and the conclusion will be different. We may say:

The consumer is rich in proportion as he *buys* at a low price. He buys at a low price in proportion to the abundance of the articles in demand; *abundance,* then, enriches him. This reasoning, extended to all consumers, must lead to the *theory of abundance.*

Which theory is right?

Can we hesitate to say? Suppose that by following out the *scarcity theory,* suppose that through prohibitions and restrictions we were compelled not only to make our own iron, but to grow our own coffee; in short, to obtain everything with difficulty and great outlay of labor. We then take an account of stock and see what our savings are.

Afterward, to test the other theory, suppose we remove the duties on iron, the duties on coffee, and the duties on everything else, so that we shall obtain everything with as little difficulty and outlay of labor as possible. If we then take an account of stock, is it not certain that we shall find more iron in the country, more coffee, more everything else?

Choose then, fellow-countrymen, between scarcity and abundance, between much and little, between Protection and Free Trade. You now know which theory is the right one, for you know the fruits they each bear.

But, it will be answered, if we are inundated with foreign goods and produce, our specie, our precious product of California, our dollars, will leave the country.

Well, what of that? Man is not fed with coin. He does not dress in gold, nor warm himself with silver. What does it matter, then, whether there be more or less specie in the country, provided there be more bread in the cupboard, more meat in the larder, more clothes in the wardrobe, and more fuel in the cellar?

Again, it will be objected, if we accustom ourselves to depend upon England for iron, what shall we do in case of a war with that country?

To this I reply, we shall then be compelled to produce iron ourselves. But, again I am told, we will not be prepared; we will have no furnaces in blast, no forges ready. True; neither will there be any time when war shall occur that the country will not be already filled with all the iron we shall want until we can make it here. Did the Confederates in the late war lack for iron? Why, then, shall we manufacture our own staples and bolts because we may some day or other have a quarrel with our ironmonger!

To sum up:

A radical antagonism exists between the vender and the buyer.

The former wishes the article offered to be *scarce,* and the supply to be small, so that the price may be high.

The latter wishes it *abundant* and the supply to be large, so that the price may be low.

The laws, which should at least remain neutral, take part for the vender against the buyer; for the producer against the consumer; for high against low prices; for scarcity against abundance; for protection against free trade. They act, if not intentionally, at least logically, upon the principle that *a nation is rich in proportion as it is in want of everything.*

Chapter 2

OBSTACLES TO WEALTH AND CAUSES OF WEALTH

Man is naturally in a state of entire destitution.

Between this state, and the satisfying of his wants, there exist a number of obstacles which it is the object of labor to surmount.

I wish to make a journey of some hundred miles. But between the point of my departure and my destination there are interposed mountains, rivers, swamps, forests, robbers; in a word—*obstacles*. To overcome these obstacles it is necessary that I should bestow much labor and great efforts in opposing them; or, what is the same thing, if others do it for me, I must pay them the value of their exertions. *It is evident that I would have been better off had these obstacles never existed.* Remember this.

Through the journey of life, in the long series of days from the cradle to the tomb, man has many difficulties to oppose him. Hunger, thirst, sickness, heat, cold, are so many obstacles scattered along his road. In a state of isolation he would be obliged to combat them all by hunting, fishing, agriculture, spinning, weaving, architecture, etc., and it is very evident that it would be better for him that these difficulties should exist to a less degree, or even not at all. In a state of society he is not obliged personally to struggle with each of these obstacles, but others do it for him; and he, in turn, must remove some one of them for the benefit of his fellow-men. This doing one kind of labor for another, is called the division of labor.

Considering mankind as a whole, *let us remember once more that it would be better for society that these obstacles should be as weak and as few as possible.*

But mark how, in viewing this simple truth from a narrow point of view, we come to believe that obstacles, instead of being a disadvantage, are actually a source of wealth!

If we examine closely and in detail the phenomena of society and the private interests of men *as modified by the division of labor,* we perceive, without difficulty, how it has happened that wants have been confounded with riches, and the obstacle with the cause.

The separation of occupations, which results from the division of labor, causes each man, instead of struggling against *all* surrounding obstacles, to combat only *one;* the effort being made not for himself alone, but for the benefit of his fellows, who, in their turn, render a similar service to him.

It hence results that this man looks upon the obstacle which he has made it his profession to combat for the benefit of others, as the immediate cause of his riches. The greater, the more serious, the more stringent, may be this obstacle, the more he is remunerated for the conquering of it, by those who are relieved by his labors.

A physician, for instance, does not busy himself in baking his bread, or in manufacturing his clothing and his instruments; others do it for him, and he, in return, combats the maladies with which his patients are afflicted. The more dangerous and frequent these maladies are, the more others are willing, the more, even, are they forced, to work in his service. Disease, then, which is an obstacle to the happiness of mankind, becomes to him the source of his comforts. The reasoning of all producers is, in what concerns themselves, the same. As the doctor draws his profits from *disease,* so does the ship-owner from the obstacle called *distance;* the agriculturist from that named *hunger;* the cloth manufacturer from *cold;* the schoolmaster lives upon *ignorance,* the jeweler upon *vanity,* the lawyer upon *cupidity and breach of faith.* Each profession has then an immediate interest in the continuation, even in the extension, of the particular obstacle to which its attention has been directed.

Theorists hence go on to found a system upon these individual interests, and say: Wants are riches: Labor is riches: The obstacle to well-being is well-being: To multiply obstacles is to give food to industry.

Then comes the statesman; and as the developing and propagating of obstacles is the developing and propagating of riches, what more natural than that he should bend his efforts to that point? He says, for instance: If we prevent a large importation of iron, we create a difficulty in procuring it. This obstacle severely felt, obliges individuals to pay, in order to relieve themselves from it. A certain number of our citizens, giving themselves up to the combating of this obstacle, will thereby make their fortunes. In proportion, too, as the obstacle is great, and the mineral scarce, inaccessible, and of difficult and distant transportation, in the same proportion will be the number of laborers maintained by the various branches of this industry.

The same reasoning will lead to the proscription of machinery.

Here are men who are at a loss how to dispose of their petroleum. This is an obstacle which other men set about removing for them by the manufacture of casks. It is fortunate, say our statesmen, that this obstacle exists, since it occupies a portion of the labor of the nation, and enriches a certain number of our citizens. But here is presented to us an ingenious machine, which cuts down the oak, squares it, makes it into staves, and, gathering these together, forms them into casks. The obstacle is thus diminished, and with it the fortunes of the coopers. We must prevent this. Let us proscribe the machine!

To sift thoroughly this sophism, it is sufficient to remember that human labor is not an *end* but a *means*.

Labor is never without employment. If one obstacle is removed, it seizes another, and mankind is delivered from two obstacles by the same effort which was at first necessary for one. If the labor of coopers could become useless, it must take another direction. To maintain that human labor can end by wanting employment, it would be necessary to prove that mankind will cease to encounter obstacles.

Chapter 3

EFFORT—RESULT

We have seen that between our wants and their gratification many obstacles are interposed. We conquer or weaken these by the employment of our faculties. It may be said, in general terms, that industry is an effort followed by a result.

But by what do we measure our well-being? By our riches? By the result of our effort, or by the effort itself? There exists always a proportion between the effort employed and the result obtained. Does progress consist in the relative increase of the second or of the first term of this proportion—between effort or result?

Both propositions have been sustained, and in political economy opinions are divided between them.

According to the first system, riches are the result of labor. They increase in the same ratio as *the result does to the effort.* Absolute perfection, of which God is the type, consists in the infinite distance between these two terms in this relation, viz., effort none, result infinite.

The second system maintains that it is the effort itself which forms the measure of, and constitutes, our riches. Progression is the increase of the *proportion of the effect to the result.* Its ideal extreme may be represented by the eternal and fruitless efforts of Sisyphus.[1]

---

1.   We will therefore beg the reader to allow us in future, for the sake of conciseness, to designate this system under the term of *Sisyphism,* from Sisyphus, who, in punishment of his crimes, was compelled to roll a stone up hill, which fell to the bottom as fast as he rolled it to the top, so that his labor was interminable as well as fruitless.

The first system tends naturally to the encouragement of everything which diminishes difficulties, and augments production—as powerful machinery, which adds to the strength of man; the exchange of produce, which allows us to profit by the various natural agents distributed in different degrees over the surface of our globe; the intellect which discovers, the experience which proves, and the emulation which excites.

The second as logically inclines to everything which can augment the difficulty and diminish the product; as, privileges, monopolies, restrictions, prohibition, suppression of machinery, sterility, &c.

It is well to mark here that the universal practice of men is always guided by the principle of the first system. Every *workman,* whether agriculturist, manufacturer, merchant, soldier, writer or philosopher, devotes the strength of his intellect to do better, to do more quickly, more economically—in a word, *to do more with less.*

The opposite doctrine is in use with theorists, essayists, statesmen, ministers, men whose business is to make experiments upon society. And even of these we may observe, that in what personally concerns themselves, they act, like everybody else, upon the principle of obtaining from their labor the greatest possible quantity of useful results.

It may be supposed that I exaggerate, and that there are no true Sisyphists.

I grant that in practice the principle is not pushed to its extreme consequences. And this must always be the case when one starts upon a wrong principle, because the absurd and injurious results to which it leads, cannot but check it in its progress. For this reason, practical industry never can admit of Sisyphism. The error is too quickly followed by its punishment to remain concealed. But in the speculative industry of theorists and statesmen, a false principle may be for a long time followed up, before the complication of its consequences, only half understood, can prove its falsity; and even when all is revealed, the opposite principle is acted upon, self is contradicted, and justification sought, in the incomparably absurd modern axiom, that in political economy there is no principle universally true.

Let us see, then, if the two opposite principles I have laid down do not predominate, each in its turn; the one in practical industry, the other in industrial legislation. When a man prefers a good plough to a bad one;

when he improves the quality of his manures; when, to loosen his soil, he substitutes as much as possible the action of the atmosphere for that of the hoe or the harrow; when he calls to his aid every improvement that science and experience have revealed, he has, and can have, but one object, viz., to *diminish the proportion of the effort to the result.* We have indeed no other means of judging of the success of an agriculturist or of the merits of his system, but by observing how far he has succeeded in lessening the one, while he increases the other; and as all the farmers in the world act upon this principle, we may say that all mankind are seeking, no doubt for their own advantage, to obtain at the lowest price, bread, or whatever other article of produce they may need, always diminishing the effort necessary for obtaining any given quantity thereof.

This incontestable tendency of human nature, once proved, would, one might suppose, be sufficient to point out the true principle to the legislator, and to show him how he ought to assist industry (if indeed it is any part of his business to assist it at all), for it would be absurd to say that the laws of men should operate in an inverse ratio from those of Providence.

Yet we have heard members of Congress exclaim, "I do not understand this theory of cheapness; I would rather see bread dear, and work more abundant." And consequently these gentlemen vote in favor of legislative measures whose effect is to shackle and impede commerce, precisely because by so doing we are prevented from procuring indirectly, and at low price, what direct production can only furnish more expensively.

Restriction has for its avowed object and acknowledged effect, the augmentation of labor. And again, equally avowed and acknowledged, its object and effect are, the increase of prices—a synonymous term for scarcity of produce. Pushed then to its greatest extreme, it is pure Sisyphism as we have defined it; *labor infinite; result nothing.* Now it is very evident that the system of Mr. So-and-so, the Congressman, is directly opposed to that of Mr. So-and-so, the agriculturist. Were he consistent with himself, he would as legislator vote against all restriction; or else as farmer, he would practise in his fields the same principle which he proclaims in the public councils. We would then see him sowing his grain in his most sterile fields, because he would thus succeed in *laboring much,* to *obtain little.* We would see him forbidding the use of the plough, because he could, by scratching up the soil with his nails, fully gratify his double wish of "*dear bread* and *abundant labor.*"

There have been men who accused railways of *injuring shipping;* and it is certainly true that the most perfect means of attaining an object must always limit the use of a less perfect means. But railways can only injure shipping by drawing from it articles of transportation; this they can only do by transporting more cheaply; and they can only transport more cheaply, by *diminishing the proportion of the effort employed to the result obtained*—for it is in this that cheapness consists. When, therefore, these men lament the suppression of labor in attaining a given result, they maintain the doctrine of Sisyphism. Logically, if they prefer the vessel to the railway, they should also prefer the wagon to the vessel, the pack-saddle to the wagon, and the sack to the pack-saddle: for this is, of all known means of transportation, the one which requires the greatest amount of labor, in proportion to the result obtained.

"Labor constitutes the riches of the people," say some theorists. This was no elliptical expression, meaning that the "results of labor constitute the riches of the people." No; these theorists intended to say, that it is the *intensity* of labor which measures riches; and the proof of this is that from step to step, from restriction to restriction, they forced on the United States (and in so doing believed that they were doing well) to give to the procuring of, for instance, a certain quantity of iron, double the necessary labor. In England, iron was then at $20; in the United States it cost $40. Supposing the day's work to be worth $2.50, it is evident that the United States could, by barter, procure a ton of iron by eight days' labor taken from the labor of the nation. Thanks to the restrictive measures of these gentlemen, sixteen days' work were necessary to procure it, by direct production. Here then we have double labor for an identical result; therefore double riches; and riches, measured not by the result, but by the intensity of labor. Is not this pure and unadulterated Sisyphism?

That there may be nothing equivocal, these gentlemen carry their idea still farther, and on the same principle that we have heard them call the intensity of labor *riches,* we will find them calling the abundant results of labor and the plenty of everything proper to the satisfying of our wants, *poverty.* "Everywhere," they remark, "machinery has pushed aside manual labor; everywhere production is superabundant; everywhere the equilibrium is destroyed between the power of production and that of consumption." Here then we see that, according to these gentlemen, if the United States was in a critical situation it was because her productions were too abundant; there was too much intelligence, too much efficiency in her national labor. We were too well fed, too well clothed, too well supplied with everything; the rapid production was more than sufficient for our wants. It was necessary to put an end to this calamity, and therefore it became needful to force us, by restrictions, to work more in order to produce less.

All that we could have further to hope for, would be, that human intellect might sink and become extinct; for, while intellect exists, it cannot but seek continually to increase the *proportion of the end to the means; of the product to the labor.* Indeed it is in this continuous effort, and in this alone, that intellect consists.

Sisyphism has been the doctrine of all those who have been intrusted with the regulation of the industry of our country. It would not be just to reproach them with this; for this principle becomes that of our administration only because it prevails in Congress; it prevails in Congress only because it is sent there by the voters; and the voters are imbued with it only because public opinion is filled with it to repletion.

Let me repeat here, that I do not accuse the protectionists in Congress of being absolutely and always Sisyphists. Very certainly they are not such in their personal transactions; very certainly each of them will procure for himself *by barter,* what by *direct production* would be attainable only at a higher price. But I maintain that they are Sisyphists when they prevent the country from acting upon the same principle.

Chapter 4

EQUALIZING OF THE FACILITIES OF PRODUCTION

The protectionists often use the following argument:

"It is our belief that protection should correspond to, should be the representation of, the difference which exists between the price of an article of home production and a similar article of foreign production. A protective duty calculated upon such a basis does nothing more than secure free competition; free competition can only exist where there is an equality in the facilities of production. In a horse-race the load which each horse carries is weighed and all advantages equalized; otherwise there could be no competition. In commerce, if one producer can undersell all others, he ceases to be a competitor and becomes a monopolist. Suppress the protection which represents the difference of price according to each, and foreign produce must immediately inundate and obtain the monopoly of our market. Every one ought to wish, for his own sake and for that of the community, that the productions of the country should be protected against foreign competition, *whenever the latter may be able to undersell the former.*"

This argument is constantly recurring in all writings of the protectionist school. It is my intention to make a careful investigation of its merits, and I must begin by soliciting the attention and the patience of the reader. I will first examine into the inequalities which depend upon natural causes, and afterwards into those which are caused by diversity of taxes.

Here, as elsewhere, we find the theorists who favor protection taking part with the producer. Let us consider the case of the unfortunate consumer, who seems to have entirely escaped their attention. They compare the field of protection to the *turf.* But on the turf, the race is at once a *means and an end.* The public has no interest in the struggle, independent of the struggle itself. When your horses are started in the course with the single object of determining which is the best runner, nothing is more natural than that their burdens should be equalized. But if your object were to send an important and critical piece of intelligence, could you without incongruity place obstacles to the speed of that one whose fleetness would secure you the best means of attaining your end? And yet this is your course in relation to industry. You forget the end aimed at, which is the *well-being* of the community; you set it aside; more, you sacrifice it by a perfect *petitio principii.*

But we cannot lead our opponents to look at things from our point of view; let us now take theirs: let us examine the question as producers.

I will seek to prove:

1. That equalizing the facilities of production is to attack the foundations of mutual exchange.

2. That it is not true that the labor of one country can be crushed by the competition of more favored climates.

3. That, even were this the case, protective duties cannot equalize the facilities of production.

4. That freedom of trade equalizes these conditions as much as possible; and

5. That the countries which are the least favored by nature are those which profit most by mutual exchange.

1. *Equalizing the facilities of production is to attack the foundations of mutual exchange.* The equalizing of the facilities of production, is not only the shackling of certain articles of commerce, but it is the attacking of the system of mutual exchange in its very foundation principle. For this system is based precisely upon the very diversities, or, if the expression be preferred, upon the inequalities of fertility, climate, temperature, capabilities, which the protectionists seek to render null. If New England sends its manufactures to the West, and the West sends corn to New England, it is because these two sections are, from different circumstances, induced to turn their attention to the production of different articles. Is there any other rule for international exchanges?

Again, to bring against such exchanges the very inequalities of condition which excite and explain them, is to attack them in their very cause of being. The protective system, closely followed up, would bring men to live like snails, in a state of complete isolation. In short, there is not one of its sophisms, which, if carried through by vigorous deductions, would not end in destruction and annihilation.

2. *It is not true that the labor of one country can be crushed by the competition of more favored climates.* The statement is not true that the unequal facility of production, between two similar branches of industry, should necessarily cause the destruction of the one which is the least fortunate. On the turf, if one horse gains the prize, the other loses it; but when two horses work to produce any useful article, each produces in proportion to his strength; and because the stronger is the more useful it does not follow that the weaker is good for nothing. Wheat is cultivated in every section of the United States, although there are great differences in the degree of fertility existing among them. If it happens that there be one which does not cultivate it, it is because, even to itself, such cultivation is not useful. Analogy will show us, that under the influences of an unshackled trade, notwithstanding similar differences, wheat would be produced in every portion of the world; and if any nation were induced to entirely abandon the cultivation of it, this would only be because it would *be her interest* to otherwise employ her lands, her capital, and her labor. And why does not the fertility of one department paralyze the agriculture of a neighboring and less favored one? Because the phenomena of political economy have a suppleness, an elasticity, and, so to speak, *a self-levelling power,* which seems to escape the attention of the school of protectionists. They accuse us of being theoretic, but it is themselves who are so to a supreme degree, if the being theoretic consists in building up systems upon the experience of a single fact, instead of profiting by the experience of a series of facts. In the above example, it is the difference in the value of lands which compensates for the difference in their fertility. Your field produces three times as much as mine. Yes. But it has cost you ten times as much, and therefore I can still compete with you: this is the sole mystery. And observe how the advantage on one point leads to disadvantage on the other. Precisely because your soil is more fruitful it is more dear. It is not *accidentally* but *necessarily* that the equilibrium is established, or at least inclines to establish

itself: and can it be denied that perfect freedom in exchanges is of all systems the one which favors this tendency?

I have cited an agricultural example; I might as easily have taken one from any trade. There are tailors at Barnegat, but that does not prevent tailors from being in New York also, although the latter have to pay a much higher rent, as well as higher price for furniture, workmen, and food. But their customers are sufficiently numerous not only to reestablish the balance, but also to make it lean on their side.

When, therefore, the question is about equalizing the advantages of labor, it would be well to consider whether the natural freedom of exchange is not the best umpire.

This self-levelling faculty of political phenomena is so important, and at the same time so well calculated to cause us to admire the providential wisdom which presides over the equalizing government of society, that I must ask permission a little longer to turn to it the attention of the reader.

The protectionists say, Such a nation has the advantage over us, in being able to procure cheaply, coal, iron, machinery, capital; it is impossible for us to compete with it.

We must examine this proposition under other aspects. For the present, I stop at the question, whether, when an advantage and a disadvantage are placed in juxtaposition, they do not bear in themselves, the former a descending, the latter an ascending power, which must end by placing them in a just equilibrium?

Let us suppose the countries A and B. A has every advantage over B; you thence conclude that labor will be concentrated upon A, while B must be abandoned. A, you say, sells much more than it buys; B buys much more than it sells. I might dispute this, but I will meet you upon your own ground.

In the hypothesis, labor being in great demand in A, soon rises in value; while labor, iron, coal, lands, food, capital, all being little sought after in B, soon fall in price.

Again: A being always selling and B always buying, cash passes from B to A. It is abundant in A, very scarce in B.

But where there is abundance of cash, it follows that in all purchases a large proportion of it will be needed. Then in A, *real dearness,* which proceeds from a very active demand, is added to *nominal dearness,* the consequence of a superabundance of the precious metals.

Scarcity of money implies that little is necessary for each purchase. Then in B, a *nominal cheapness* is combined with *real cheapness.*

Under these circumstances, industry will have the strongest possible motives for deserting A to establish itself in B.

Now, to return to what would be the true course of things. As the progress of such events is always gradual, industry from its nature being opposed to sudden transits, let us suppose that, without waiting the extreme point, it will have gradually divided itself between A and B, according to the laws of supply and demand; that is to say, according to the laws of justice and usefulness.

I do not advance an empty hypothesis when I say, that were it possible that industry should concentrate itself upon a single point, there must, from its nature, arise spontaneously, and in its midst, AN IRRESISTIBLE POWER OF DECENTRALIZATION.

We will quote the words of a manufacturer to the Chamber of Commerce at Manchester (the figures brought into his demonstration being suppressed):

"Formerly we exported goods; this exportation gave way to that of thread for the manufacture of goods; later, instead of thread, we exported machinery for the making of thread; then capital for the construction of machinery; and lastly, workmen and talent, which are the source of capital. All these elements of labor have, one after the other, transferred themselves to other points, where their profits were increased, and where the means of subsistence being less difficult to obtain, life is maintained at less cost. There are at present to be seen in Prussia, Austria, Saxony, Switzerland, and Italy, immense manufacturing establishments, founded entirely by English capital, worked by English labor, and directed by English talent."

We may here perceive that Nature, with more wisdom and foresight than the narrow and rigid system of the protectionists can suppose, does not permit the concentration of labor, and the monopoly of advantages, from which they draw their arguments as from an absolute and irremediable fact. It has, by means as simple as they are infallible, provided for dispersion, diffusion, mutual dependence, and simultaneous progress; all of which, your restrictive laws paralyze as much as is in their power, by their tendency towards the isolation of nations. By this means they render much more decided the differences existing in the conditions of production; they check the self-levelling power of industry, prevent fusion of interests, neutralize the counterpoise, and fence in each nation within its own peculiar advantages and disadvantages.

3. *Even were the labor of one country crushed by the competition of more favored climates (which is denied), protective duties cannot equalize the facilities of production.* To say that by a protective law the conditions of production are equalized, is to disguise an error under false terms. It is not true that an import duty equalizes the conditions of production. These remain after the imposition of the duty just as they were before. The most that law can do is to equalize the *conditions of sale*. If it should be said that I am playing upon words, I retort the accusation upon my adversaries. It is for them to prove that *production* and *sale* are synonymous terms, which if they cannot do, I have a right to accuse them, if not of playing upon words, at least of confounding them.

Let me be permitted to exemplify my idea.

Suppose that several New York speculators should determine to devote themselves to the production of oranges. They know that the oranges of Portugal can be sold in New York at one cent each, whilst on account of the boxes, hot-houses, &c., which are necessary to ward against the severity of our climate, it is impossible to raise them at less than a dollar apiece. They accordingly demand a duty of ninety-nine cents upon Portugal oranges. With the help of this duty, say they, the *conditions of production* will be equalized. Congress, yielding as usual to this argument, imposes a duty of ninety-nine cents on each foreign orange.

Now I say that the *relative conditions of production* are in no wise changed. The law can take nothing from the heat of the sun in Lisbon, nor from the severity of the frosts in New York. Oranges continuing to mature themselves *naturally* on the banks of the Tagus, and artificially upon those of the Hudson, must continue to require for their production much more labor on the latter than the former. The law can only equalize the *conditions of sale*. It is evident that while the Portuguese sell their oranges here at a dollar apiece, the ninety-nine cents which go to pay the tax are taken from the American consumer. Now look at the whimsicality of the result. Upon each Portuguese orange, the country loses nothing; for the ninety-nine cents which the consumer pays to satisfy the impost tax, enter into the treasury. There is improper distribution; but no loss. But upon each American orange consumed, there will be about ninety-nine cents lost; for while the buyer very certainly loses them, the seller just as certainly does not gain them; for, even according to the hypothesis, he will receive only the price of production, I will leave it to the protectionists to draw their conclusion.

4. *But freedom of trade equalizes these conditions as much as is possible.* I have laid some stress upon this distinction between the conditions of production and those of sale, which perhaps the prohibitionists may consider as paradoxical, because it leads me on to what they will consider as a still stranger paradox. This is: If you really wish to equalize the facilities of production, leave trade free.

This may surprise the protectionists; but let me entreat them to listen, if it be only through curiosity, to the end of my argument. It shall not be long. I will now take it up where we left off.

I will add that free trade equalizes also the facilities for attaining enjoyments, comforts, and general consumption; the last, an object which is, it would seem, quite forgotten, and which is nevertheless all-important; since, in fine, consumption is the main object of all our industrial efforts. Thanks to freedom of trade, we would enjoy here the results of the Portuguese sun, as well as Portugal itself; and the inhabitants of New York would have in their reach, as well as those of London, and with the same facilities, the advantages which nature has in a mineralogical point of view conferred upon Cornwall.If we suppose for the moment, that the common and daily profits of each American amount to one dollar, it will indisputably follow that to produce an orange by *direct* labor in America, one day's work, or its equivalent, will be requisite; whilst to produce the cost of a Portuguese orange, only one-hundredth of this day's labor is required; which means simply this, that the sun does at Lisbon what labor does at New York. Now is it not evident, that if I can produce an orange, or, what is the same thing, the means of buying it, with one-hundredth of a day's labor, I am placed exactly in the same condition as the Portuguese producer himself, excepting the expense of the transportation? It therefore follows that freedom of commerce equalizes the conditions of production direct or indirect, as much as it is possible to equalize them; for it leaves but the one inevitable difference, that of transportation.

5. *Countries least favored by nature (countries not yet cleared of forests, for example) are those which profit most by mutual exchange.* The protectionists may suppose me in a paradoxical humor, for I go further still. I say, and I sincerely believe, that if any two countries are placed in unequal circumstances as to advantages of production, *the one of the two which is the less favored by nature, will gain more by freedom of commerce.* To prove this, I will be obliged to turn somewhat aside from the form of reasoning which belongs to this work. I will do so, however; first, because the question in discussion turns upon this point; and again, because it will give me the opportunity of exhibiting a law of political economy of the highest importance, and which, well understood, seems to me to be destined to lead back to this science all those sects which, in our days, are seeking in the land of chimeras that social harmony which they have been unable to discover in nature. I speak of the law of consumption, which the majority of political economists may well be reproached with having too much neglected.

Consumption is the *end,* the final cause of all the phenomena of political economy, and, consequently, in it is found their final solution.

No effect, whether favorable or unfavorable, can be vested permanently in the producer. His advantages and disadvantages, derived from his relations to nature and to society, both pass gradually from him; and by an almost insensible tendency are absorbed and fused into the community at large—the community considered as consumers. This is an admirable law, alike in its cause and its effects; and he who shall succeed in making it well understood, will have a right to say, "I have not, in my passage through the world, forgotten to pay my tribute to society."

Every circumstance which favors the work of production is of course hailed with joy by the producer, for its *immediate effect* is to enable him to render greater services to the community, and to exact from it a greater remuneration. Every circumstance which injures production, must equally be the source of uneasiness to him; for its *immediate effect* is to diminish his services, and consequently his remuneration. This is a fortunate and necessary law of nature. The immediate good or evil of favorable or unfavorable circumstances must fall upon the producer, in order to influence him invisibly to seek the one and to avoid the other.

Again: when an inventor succeeds in his labor-saving machine, the *immediate* benefit of this success is received by him. This again is necessary, to determine him to devote his attention to it. It is also just; because it is just that an effort crowned with success should bring its own reward.

But these effects, good and bad, although permanent in themselves, are not so as regards the producer. If they had been so, a principle of progressive and consequently infinite inequality would have been introduced among men. This good, and this evil, both therefore pass on, to become absorbed in the general destinies of humanity.

How does this come about? I will try to make it understood by some examples.

Let us go back to the thirteenth century. Men who gave themselves up to the business of copying, received for this service *a remuneration regulated by the general rate of the profits*. Among them is found one, who seeks and finds the means of rapidly multiplying copies of the same work. He invents printing. The first effect of this is, that the individual is enriched, while many more are impoverished. At the first view, wonderful as the discovery is, one hesitates in deciding whether it is not more injurious than useful. It seems to have introduced into the world, as I said above, an element of infinite inequality. Guttenberg makes large profits by this invention, and perfects the invention by the profits, until all other copyists are ruined. As for the public—the consumer—it gains but little, for Guttenberg takes care to lower the price of books only just so much as is necessary to undersell all rivals.

But the great Mind which put harmony into the movements of celestial bodies, could also give it to the internal mechanism of society. We will see the advantages of this invention escaping from the individual, to become for ever the common patrimony of mankind.

The process finally becomes known. Guttenberg is no longer alone in his art; others imitate him. Their profits are at first considerable. They are recompensed for being the first who made the effort to imitate the processes of the newly-invented art. This again was necessary, in order that they might be induced to the effort, and thus forward the great and final result to which we approach. They gain largely; but they gain less than the inventor, for *competition* has commenced its work. The price of books now continually decreases. The gains of the imitators diminish in proportion as the invention becomes older; and in the same proportion imitation becomes less meritorious. Soon the new object of industry attains its normal condition; in other words, the remuneration of printers is no longer an exception to the general rules of remuneration, and, like that of copyists formerly, it is only regulated *by the general rate of profits.* Here then the producer, as such, holds only the old position. The discovery, however, has been made; the saving of time, labor, effort, for a fixed result, for a certain number of volumes, is realized. But in what is this manifested? In the cheap price of books. For the good of whom? For the good of the consumer—of society—of humanity. Printers, having no longer any peculiar merit, receive no longer a peculiar remuneration. As men—as consumers—they no doubt participate in the advantages which the invention confers upon the community; but that is all. As printers, as producers, they are placed upon the ordinary footing of all other producers. Society pays them for their labor, and not for the usefulness of the invention. *That* has become a gratuitous benefit, a common heritage to mankind.

The wisdom and beauty of these laws strike me with admiration and reverence.

What has been said of printing, can be extended to every agent for the advancement of labor—from the nail and the mallet, up to the locomotive and the electric telegraph. Society enjoys all, by the abundance of its use, its consumption; and it *enjoys all gratuitously.* For as their effect is to diminish prices, it is evident that just so much of the price as is taken off by their intervention, renders the production in so far *gratuitous.* There only remains the actual labor of man to be paid for; and the remainder, which is the result of the invention, is subtracted; at least after the invention has run through the cycle which I have just described as its destined course. I send for a workman; he brings a saw with him; I pay him two dollars for his day's labor, and he saws me twenty-five boards. If the saw had not been invented, he would perhaps not have been able to make one board, and I would none the less have paid him for his day's labor. The *usefulness,* then, of the saw, is for me a gratuitous gift of nature, or rather, is a portion of the inheritance which, *in common* with my brother men, I have received from the genius of my ancestors. I have two workmen in my field; the one directs the handle of a plough, the other that of a spade. The result of their day's labor is very different, but the price is the same, because the remuneration is proportioned, not to the usefulness of the result, but to the effort, the [time, and] labor given to attain it.

I invoke the patience of the reader, and beg him to believe, that I have not lost sight of free trade: I entreat him only to remember the conclusion at which I have arrived: *Remuneration is not proportioned to the usefulness of the articles brought by the producer into the market, but to the [time and] labor required for their production.*[1]

I have so far taken my examples from human inventions, but will now go on to speak of natural advantages.

In every article of production, nature and man must concur. But the portion of nature is always gratuitous. Only so much of the usefulness of an article as is the result of human labor becomes the object of mutual exchange, and consequently of remuneration. The remuneration varies much, no doubt, in proportion to the intensity of the labor, of the skill, which it requires, of its being *a-propos* to the demand of the day, of the need which exists for it, of the momentary absence of competition, &c. But it is not the less true in principle, that the assistance received from natural laws, which belongs to all, counts for nothing in the price.

We do not pay for the air we breathe, although so useful to us, that we could not live two minutes without it. We do not pay for it, because nature furnishes it without the intervention of man's labor. But if we wish to separate one of the gases which compose it for instance, to fill a balloon, we must take some [time and] labor; or if another takes it for us, we must give him an equivalent in something which will have cost us the trouble of production. From which we see that the exchange is between efforts, [time and] labor. It is certainly not for hydrogen gas that I pay, for this is everywhere at my disposal, but for the work that it has been necessary to accomplish in order to disengage it; work which I have been spared, and which I must refund. If I am told that there are other things to pay for, as

---

1. It is true that [time and] labor do not receive a uniform remuneration; because labor is more or less intense, dangerous, skilful, &c., [and time more or less valuable.] Competition establishes for each category a price current: and it is of this variable price that I speak.

expense, materials, apparatus, I answer, that still in these things it is the work that I pay for. The price of the coal employed is only the representation of the [time and] labor necessary to dig and transport it.

We do not pay for the light of the sun, because nature alone gives it to us. But we pay for the light of gas, tallow, oil, wax, because here is labor to be remunerated;—and remark, that it is so entirely [time and] labor and not utility to which remuneration is proportioned, that it may well happen that one of these means of lighting, while it may be much more effective than another, may still cost less. To cause this, it is only necessary that less [time and] human labor should be required to furnish it.

When the water-boat comes to supply my ship, were I to pay in proportion to the *absolute utility* of the water, my whole fortune would not be sufficient. But I pay only for the trouble taken. If more is required, I can get another boat to furnish it, or finally go and get it myself. The water itself is not the subject of the bargain, but the labor required to obtain the water. This point of view is so important, and the consequences that I am going to draw from it so clear, as regards the freedom of international exchanges, that I will still elucidate my idea by a few more examples.

The alimentary substance contained in potatoes does not cost us very dear, because a great deal of it is attainable with little work. We pay more for wheat, because, to produce it, Nature requires more labor from man. It is evident that if Nature did for the latter what she does for the former, their prices would tend to the same level. It is impossible that the producer of wheat should permanently gain more than the producer of potatoes. The law of competition cannot allow it.

Again, if by a happy miracle the fertility of all arable lands were to be increased, it would not be the agriculturist, but the consumer, who would profit by this phenomenon; for the result of it would be abundance and cheapness. There would be less labor incorporated into an acre of grain, and the agriculturist would be therefore obliged to exchange it for less labor incorporated into some other article. If, on the contrary, the fertility of the soil were suddenly to deteriorate, the share of nature in production would be less, that of labor greater, and the result would be higher prices.

I am right then in saying that it is in consumption, in mankind, that at length all political phenomena find their solution. As long as we fail to follow their effects to this point, and look only at *immediate* effects, which act but upon individual men or classes of men *as producers,* we know nothing more of political economy than the quack does of medicine, when instead of following the effects of a prescription in its action upon the whole system, he satisfies himself with knowing how it affects the palate and the throat.

The tropical regions are very favorable to the production of sugar and coffee; that is to say, Nature does most of the business and leaves but little for labor to accomplish. But who reaps the advantage of this liberality of Nature? *Not these regions,* for they are forced by competition to receive remuneration simply for their labor. It is *mankind* who is the gainer; for the result of this liberality is *cheapness,* and cheapness belongs to the world.

Here in the temperate zone, we find coal and iron ore on the surface of the soil; we have but to stoop and take them. At first, I grant, the immediate inhabitants profit by this fortunate circumstance. But soon comes competition, and the price of coal and iron falls, until this gift of nature becomes gratuitous to all, and human labor is only paid according to the general rate of profits.

Thus, natural advantages, like improvements in the process of production, are, or have, a constant tendency to become, under the law of competition, the common and *gratuitous* patrimony of consumers, of society, of mankind. Countries, therefore, which do not enjoy these advantages, must gain by commerce with those which do; because the exchanges of commerce are between *labor and labor,* subtraction being made of all the natural advantages which are combined with these labors; and it is evidently the most favored countries which can incorporate into a given labor the largest proportion of these *natural advantages.* Their produce representing less labor, receives less recompense; in other words, is *cheaper.* If then all the liberality of Nature results in cheapness, it is evidently not the producing, but the consuming country, which profits by her benefits.

Hence we may see the enormous absurdity of the consuming country, which rejects produce precisely because it is cheap. It is as though we should say: "We will have nothing of that which Nature gives you. You ask of us an effort equal to two, in order to furnish ourselves with produce only attainable at home by an effort equal to four. You can do it because with you Nature does half the work. But we will have nothing to do with it; we will wait till your climate, becoming more inclement, forces you to ask of us a labor equal to four, and then we can treat with you *upon an equal footing!*"

A is a favored country; B is maltreated by Nature. Mutual traffic then is advantageous to both, but principally to B, because the exchange is not between *utility* and *utility,* but between *value* and *value.* Now A furnishes a greater *utility in a similar value,* because the utility of any article includes at once what Nature and what labor have done; whereas the value of it only corresponds to the portion accomplished by labor. B then makes an entirely advantageous bargain; for by simply paying the producer from A for his labor, it receives in return not only the results of that labor, but in addition there is thrown in whatever may have accrued from the superior bounty of Nature.

We will lay down the general rule.

Traffic is an exchange of *values;* and as value is reduced by competition to the simple representation of labor, traffic is the exchange of equal labors. Whatever Nature has done towards the production of the articles exchanged, is given on both sides *gratuitously;* from whence it necessarily follows, that the most advantageous commerce is transacted with those countries which are the least favored by Nature.

The theory of which I have attempted in this chapter to trace the outlines, deserves a much greater elaboration. But perhaps the attentive reader will have perceived in it the fruitful seed which is destined in its future growth to smother Protectionism, at once with the various other isms whose object is to exclude the law of *competition* from the government of the world. Competition, no doubt, considering man as producer, must often interfere with his individual and *immediate* interests. But if we consider the great object of all labor, the universal good, in a word, Consumption, we cannot fail to find that Competition is to the moral world what the law of equilibrium is to the material one. It is the foundation of true gratification, of true Liberty and Equality, of the equality of comforts and condition, so much sought after in our day; and if so many sincere reformers, so many earnest friends to public right, seek to reach their end by *commercial legislation,* it is only because they do not yet understand *commercial freedom.*

Chapter 5

OUR PRODUCTIONS ARE OVERLOADED WITH INTERNAL TAXES

This is but a new wording of the Sophism before noticed. The demand made is, that the foreign article should be taxed, in order to neutralize the effects of the internal tax, which weighs down domestic produce. It is still then but the question of equalizing the facilities of production. We have but to say that the tax is an artificial obstacle, which has exactly the same effect as a natural obstacle, i.e. the increasing of the price. If this increase is so great that there is more loss in producing the article in question at home than in attracting it from foreign parts by the production of an equivalent value of something else—*laissez faire*. Individual interest will soon learn to choose the lesser of two evils. I might refer the reader to the preceding demonstration for an answer to this Sophism; but it is one which recurs so often, that it deserves a special discussion.

I have said more than once, that I am opposing only the theory of the protectionists, with the hope of discovering the source of their errors. Were I disposed to enter into controversy with them, I would say: Why direct your tariffs principally against England, a country more overloaded with taxes than any in the world? Have I not a right to look upon your argument as a mere pretext? But I am not of the number of those who believe that prohibitionists are guided by interest, and not by conviction. The doctrine of Protection is too popular not to be sincere. If the majority could believe in freedom, we would be free. Without doubt it is individual interest which weighs us down with tariffs; but it acts upon conviction. "The will (said Pascal) is one of the principal organs of belief." But belief does not the less exist because it is rooted in the will and in the secret inspirations of egotism.

We will return to the Sophism drawn from internal taxes.

As to unproductive taxes, suppress them if you can; but surely it is a most singular idea to suppose, that their evil effect is to be neutralized by the addition of individual taxes to public taxes. Many thanks for the compensation! The State, you say, has taxed us too much; surely this is no reason that we should tax each other!The government may make either a good or a bad use of taxes; it makes a good use of them when it renders to the public services equivalent to the value received from them; it makes a bad use of them when it expends this value, giving nothing in return. To say in the first case that they place the country which pays them in more disadvantageous conditions for production, than the country which is free from them, is a Sophism. We pay, it is true, so many millions for the administration of justice, and the maintenance of order, but we have justice and order; we have the security which they give, the time which they save for us; and it is most probable that production is neither more easy nor more active among nations, where (if there be such) each individual takes the administration of justice into his own hands. We pay, I grant, many millions for roads, bridges, ports, steamships; but we have these steamships, these ports, bridges, and roads; and unless we maintain that it is a losing business to establish them, we cannot say that they place us in a position inferior to that of nations who have, it is true, no budget of public works, but who likewise have no public works. And here we see why (even while we accuse taxes of being a cause of industrial inferiority) we direct our tariffs precisely against those nations which are the most taxed. It is because these taxes, well used, far from injuring, have ameliorated the *conditions of production* to these nations. Thus we again arrive at the conclusion that the protectionist Sophisms not only wander from, but are the contrary—the very antithesis—of truth.

A protective duty is a tax directed against foreign produce, but which returns, let us keep in mind, upon the national consumer. Is it not then a singular argument to say to him, "Because the taxes are heavy, we will raise prices higher for you; and because the State takes a part of your revenue, we will give another portion of it to benefit a monopoly?"

But let us examine more closely this Sophism so accredited among our legislators; although, strange to say, it is precisely those who keep up the unproductive taxes (according to our present hypothesis) who attribute to them afterwards our supposed inferiority, and seek to re-establish the equilibrium by further taxes and new clogs.

It appears to me to be evident that protection, without any change in its nature and effects, might have taken the form of a direct tax, raised by the State, and distributed as a premium to privileged industry.

Let us admit that foreign iron could be sold in our market at $16, but not lower; and American iron at not lower than $24.

In this hypothesis there are two ways in which the State can secure the national market to the home producer.

The first, is to put upon foreign iron a duty of $10. This, it is evident, would exclude it, because it could no longer be sold at less than $26; $16 for the indemnifying price, $10 for the tax; and at this price it must be driven from the market by American iron, which we have supposed to cost $24. In this case the buyer, the consumer, will have paid all the expenses of the protection given.

The second means would be to lay upon the public an Internal Revenue tax of $10, and to give it as a premium to the iron manufacturer. The effect would in either case be equally a protective measure. Foreign iron would, according to both systems, be alike excluded; for our iron manufacturer could sell at $14, what, with the $10 premium, would thus bring him in $24. While the price of sale being $14, foreign iron could not obtain a market at $16.

In these two systems the principle is the same; the effect is the same. There is but this single difference; in the first case the expense of protection is paid by a part, in the second by the whole of the community. I frankly confess my preference for the second system, which I regard as more just, more economical, and more legal. More just, because, if society wishes to give bounties to some of its members, the whole community ought to contribute; more economical, because it would banish many difficulties, and save the expenses of collection; more legal, because the public would see clearly into the operation, and know what was required of it.

But if the protective system had taken this form, would it not have been laughable enough to hear it said: "We pay heavy taxes for the army, the navy, the judiciary, the public works, the debt, &c. These amount to more than 200 millions. It would therefore be desirable that the State should take another 200 millions to relieve the poor iron manufacturers."

This, it must certainly be perceived, by an attentive investigation, is the result of the Sophism in question. In vain, gentlemen, are all your efforts; you cannot give money to one without taking it from another. If you are absolutely determined to exhaust the funds of the taxable community, well; but, at least, do not mock them; do not tell them, "We take from you again, in order to compensate you for what we have already taken."

It would be a too tedious undertaking to endeavor to point out all the fallacies of this Sophism. I will therefore limit myself to the consideration of it in three points.

You argue that the United States are overburdened with taxes, and deduce thence the conclusion that it is necessary to protect such and such an article of produce. But protection does not relieve us from the payment of these taxes. If, then, individuals devoting themselves to any one object of industry, should advance this demand: "We, from our participation in the payment of taxes, have our expenses of production increased, and therefore ask for a protective duty which shall raise our price of sale:" what is this but a demand on their part to be allowed to free themselves from the burden of the tax, by laying it on the rest of the community? Their object is to balance, by the increased price of their produce, the amount which they pay in taxes. Now, as the whole amount of these taxes must enter into the Treasury, and the increase of price must be paid by society, it follows that (where this protective duty is imposed) society has to bear, not only the general tax, but also that for the protection of the article in question. But, it is answered, let *everything* be protected. Firstly, this is impossible; and, again, were it possible, how could such a system give relief? *I* will pay for you, *you* will pay for me; but not the less still there remains the tax to be paid.

Thus you are the dupes of an illusion. You determine to raise taxes for the support of an army, a navy, judges, roads, &c. Afterwards you seek to disburden from its portion of the tax, first one article of industry, then another, then a third; always adding to the burden of the mass of society.

You thus only create interminable complications. If you can prove that the increase of price resulting from protection, falls upon the foreign producer, I grant something specious in your argument. But if it be true that the American people paid the tax before the passing of the protective duty, and afterwards that it has paid not only the tax but the protective duty also, truly I do not perceive wherein it has profited.

But I go much further, and maintain that the more oppressive our taxes are, the more anxiously ought we to open our ports and frontiers to foreign nations, less burdened than ourselves. And why? *In order that we may* SHARE WITH THEM, *as much as possible, the burden which we bear.* Is it not an incontestable maxim in political economy, that taxes must, in the end, fall upon the consumer? *The greater then our commerce, the greater the portion which will be reimbursed to us, of taxes incorporated in the produce which we will have sold to foreign consumers; whilst we on our part will have made to them only a lesser reimbursement, because (according to our hypothesis) their produce is less taxed than ours.*

Chapter 6

BALANCE OF TRADE

Our adversaries have adopted a system of tactics, which embarrasses us not a little. Do we prove our doctrine? They admit the truth of it in the most respectful manner. Do we attack their principles? They abandon them with the best possible grace. They only ask that our doctrine, which they acknowledge to be true, should be confined to books; and that their principles, which they allow to be false, should be established in practice. If we will give up to them the regulation of our tariffs, they will leave us triumphant in the domain of literature.

It is constantly alleged in opposition to our principles, that they are good only in theory. But, gentlemen, do you believe that merchants' books are good in practice? It does appear to me, if there is anything which can have a practical authority, when the object is to prove profit and loss, that this must be commercial accounts. We cannot suppose that all the merchants of the world, for centuries back, should have so little understood their own affairs, as to have kept their books in such a manner as to represent gains as losses, and losses as gains. Truly it would be easier to believe that our legislators are bad political economists. A merchant, one of my friends, having had two business transactions, with very different results, I have been curious to compare on this subject the accounts of the counter with those of the custom-house, interpreted by our legislators.

Mr. T dispatched from New Orleans a vessel freighted for France with cotton valued at $200,000. Such was the amount entered at the custom-house. The cargo, on its arrival at Havre, had paid ten per cent. expenses, and was liable to thirty per cent. duties, which raised its value to

$280,000. It was sold at twenty per cent. profit on its original value, which equalled $40,000, and the price of sale was $320,000, which the consignee converted into merchandise, principally Parisian goods. These goods, again, had to pay for transportation to the sea-board, insurance, commissions, &c., ten per cent.; so that when the return cargo arrived at New Orleans, its value had risen to $352,000, and it was thus entered at the custom-house. Finally, Mr. T realized again on this return cargo twenty per cent. profits, amounting to $70,400. The goods thus sold for the sum of $422,400.

If our legislators require it, I will send them an extract from the books of Mr. T. They will there see, *credited* to the account of *profit and loss,* that is to say, set down as gained, two sums; the one of $40,000, the other of $70,400, and Mr. T feels perfectly certain that, as regards these, there is no mistake in his accounts.

Now what conclusion do our Congressmen draw from the sums entered into the custom-house, in this operation? They thence learn that the United States have exported $200,000, and imported $352,000; from whence they conclude *"that she has spent, dissipated, the profits of her previous savings; that she is impoverishing herself and progressing to her ruin; and that she has squandered on a foreign nation* $152,000 *of her capital."*

Some time after this transaction, Mr. T dispatched another vessel, again freighted with national produce, to the amount of $200,000. But the vessel foundered in leaving the port, and Mr. T had only further to inscribe upon his books two little items, thus worded:

"Sundries due to X, $200,000, for purchase of divers articles dispatched by vessel N."

"Profit and loss due, to sundries, $200,000, *for final and total loss of cargo."*

In the meantime the custom-house inscribed $200,000 upon its list of *exportations,* and as there can of course be nothing to balance this entry on the list of *importations,* it hence follows that our enlightened members of Congress must see in this wreck *a clear profit* to the United States of $200,000.

We may draw hence yet another conclusion, viz.: that according to the Balance of Trade theory, the United States has an exceedingly simple manner of constantly doubling her capital. It is only necessary, to accomplish this, that she should, after entering into the custom-house her articles for exportation, cause them to be thrown into the sea. By this course, her exportations can speedily be made to equal her capital; importations will be nothing, and our gain will be, all which the ocean will have swallowed up.

You are joking, the protectionists will reply. You know that it is impossible that we should utter such absurdities. Nevertheless, I answer, you do utter them, and what is more, you give them life, you exercise them practically upon your fellow-citizens, as much, at least, as is in your power to do.

But lest even Mr. T's books may not be deemed of sufficient weight to counterbalance the convictions of the Horace Greeley school of prohibition, I shall proceed to furnish a table exhibiting various classes of commercial transactions, embracing most of the classes usually effected by importing and exporting houses, all of which may result in undoubted profits to the parties engaged in them, and to the country at large, and yet which, as they appear in the annual Commerce and Navigation Reports issued by the government, would be made to prove by Mr. Greeley that the result has in each case been a loss to the country. The sums are all stated in gold:

A, represents one hundred merchants, who shipped to London beef, boots and shoes, butter, cheese, cotton, hams and bacon, flour, Indian corn, lard, lumber, machinery, oils, pork, staves, tallow, tobacco and cigars, worth in New York, in the aggregate, ten millions of dollars, gold, but worth in London plus the cost of transportation, &c., eleven millions of dollars, gold, in bond. After being sold in London, the proceeds (eleven millions) were invested in British goods, worth eleven millions in London, but worth twelve millions in bond in New York, and plus the cost of transportation, &c. After having these goods sold in New York, a net profit of two millions was the result of the whole transaction, a profit both to the merchants and the country; yet, according to the Commerce and Navigation Returns, the exports were ten millions, and the imports eleven millions (valued at the foreign place of production as the law directs), showing, according to Mr. Greeley's solitary point of view, a loss to the country of one million.

B, owned a gold mine in Nevada, and had no capital with which to develop it. He proceeded to France, sold his mine to C for a million, which he invested in French muslin-de-laines, buttons, and glassware, worth a million in France, but worth $1,100,000 in Philadelphia, ex duty and plus transportation, &c. These sold, B netted an undoubted profit of $100,000, besides getting rid of his mine; but, according to the Commerce and Navigation Returns, the exports were nothing, and the imports $1,000,000; showing, according to Mr. Greeley's solitary point of view, a loss to the country of $1,000,000.

C, the French owner of the Nevada mine, had a million more with which to develop it. Hearing that French cloths and gloves had a good sale in Boston, he invested his million in these goods, sailed for Boston with them, sold them there in bond and plus exportation, for $1,100,000, which he at once invested in machinery, labor, &c., destined for Nevada. So far, C made a profit of $100,000, and had $2,100,000 invested in an American gold mine; but, according to the Commerce and Navigation Returns, the exports were nothing, and the imports $1,000,000; according to Mr. Greeley's solitary point of view, a loss to the country of $1,000,000.

D, had a rich uncle in Rio Janeiro who died and left him a million. D ordered this sum to be invested in hides and shipped to him at Boston. These hides were worth a million in Rio, but $1,100,000 in Natick, ex duty and plus transportation. Upon selling them D was clearly worth $1,100,000; yet, according to the Commerce and Navigation Reports, as there had been no exports, but simply $1,000,000 of imports, the transaction, from Mr. Greeley's solitary point of view, seemed a loss to the country of $1,000,000.

E, in 1850, shipped to Cuba, wagons, carts, agricultural implements, pianos and billiard-tables, worth $1,000,000 in Baltimore, but $1,100,000 in Havana, ex duty and plus transportation. These he sold, and invested the proceeds in cigars worth $1,100,000 in Havana, but in Russia, ex duty and plus transportation, $1,210,000. Disposing of these in turn, and investing the proceeds in Russian iron worth $1,210,000 in Russia, but $1,331,000 in Venezuela, ex duty and plus transportation, he shipped the iron to Venezuela, where he realized on it, investing the proceeds this time in South American products worth in Spain $1,464,100.

He sold these products in Spain, bought olive oil with the proceeds, shipped the same to Australia, where it was worth, ex duty and plus charges, $1,610,510, which sum he realized in gold, which he carried to New York in 1853. On the latter transaction he makes no profit, but barely clears his charges. Yet on the whole he has made a net gain of $610,510; but, according to the Commerce and Navigation Reports, the exports have been $1,000,000 and the imports $1,610,510, showing, from Mr. Greeley's solitary point of view, a loss to the country of $610,510. Nay more, for Mr. Greeley balances his trade accounts each year by itself, and as E's outward shipment was made in 1850 and his importation in 1853, the country, according to H.G., lost in 1853, by over importation, $1,610,500. Yet not to be hard on H.G., and to be perfectly honest in our accounts, we will only set down a loss to the country from his point of view of $610,510.

F, owned the 4,000 ton ship Great Republic, which cost him $160,000. Finding her too large for profitable employment, and hearing that large vessels were in demand in England as troop transports to the Crimea, he sent her out in ballast and sold her in Southampton for $200,000 cash. With this sum he went to Geneva, where he invested it in Swiss watches worth $200,000 in Geneva, but $210,000 in New Orleans, ex duty and plus transportation. To New Orleans he accordingly shipped the watches, and they were sold. By these transactions he not only got rid of his elephant, but both he and the country clearly gained $50,000. Yet according to Mr. Greeley's single eye the country suffered to the extent of $200,000, for in the exports appeared nothing, but among the imports $200,000 worth of foreign gewgaws, only fit to keep time with.

G, (an actual transaction) shipped by the Great Eastern on her last voyage from New York, lard and other merchandise, worth in New York $600,000, the fact of which, in the hurry of business, he failed to report to the Custom House, and it therefore did not appear in the exports. This lard was carried to England, where it found no sale, and was reshipped to New York. G only escaped being charged duty on it when it arrived, by swearing that it had been originally shipped from here in good faith; yet it was entered as an import (free of duty), and showed, according to Mr. Greeley's one eye, that the country was on the road to ruin $600,000 worth.

H, lived in Brownsville, Texas, where he had a lot of arms and gunpowder, worth $100,000. The Mexicans levied a very high import duty on these articles, and they consequently bore a very high price in Matamoras, just opposite, being worth in the market of that town no less than $250,000. He accordingly conceived the idea of smuggling them into Mexican territory, and, with the connivance of the Mexican officials, (what rascals these foreign custom-house officials are, to be sure!) actually succeeded in doing so, and thus realized the very handsome profit of $150,000 in gold. The entire proceeds he invested in Mexican indigo and cochineal, worth in Mexico $250,000, and in Boston $275,000, in bond, plus charges. Of course, no export entry was furnished to the customs collector at Brownsville; but Mr. Greeley fastened his one eye on the indigo and cochineal, when it arrived in Boston, and made up his mind that the country had lost $250,000. As for H, he has invested $100,000 in more gunpowder and arms, and starts for Brownsville next week, to try his luck again. With the other $175,000 he has a notion of buying out the New York *Tribune,* and setting it right on free trade, and other matters of the sort.

J, had $2,000,000 in five-twenty bonds, which cost him $1,400,000 gold. As the market price in New York was only 70 gold, while it was 72–1/4 in London, he conceived the inhuman idea of selling them in the latter place. The cost of sending them there, including insurance, &c., made them net him but 72, but at this price he gained a profit of $40,000. With his capital now augmented to $1,440,000 he bought rags in Italy, which he sold in New York for $1,584,000, ex duty and plus transportation, a clear profit of $184,000 from the start. No export appearing in the Commerce and Navigation Returns, and nothing but the rags meeting his unital gaze, Mr. Greeley at once posted his national ledger with a loss of $1,440,000, the cost of the rags in Italy. I, and his friends owned a fine fleet of merchantmen when the war broke out. The aggregate burden of the vessels was nearly a million of tons, and they were worth $40 a ton. When the rebel cruisers commenced their operations, there were no United States cruisers prepared to capture them, because our best vessels were on blockade service. This being the case, insurance on American merchantmen rose very high—so high that I and his friends were reluctantly compelled to sell their vessels in Great Britain and elsewhere, and convert them into cash. They brought $40,000,000, and this sum was invested in merchandise, which netted a profit of ten per cent. to I and his friends. They thus gained $4,000,000 by these transactions. The entire proceeds, $44,000,000, they then lent to the government with which to carry on its war of existence with the Southern insurgents. Profitable as these transactions clearly were to I and his friends, and to the government, Mr. Greeley, nevertheless, only sees the import of $40,000,000 worth of foreign extravagances, and consequently wants the tariff on iron increased in order to make water run up hill.

K, was, and is still (for these are actual transactions taken from his account books), an exchange broker, doing business in New York. He buys notes on the banks of England, Ireland, Scotland, France and Canada—indeed, foreign banknotes of all kinds—for which he usually pays about ninety per cent, of their face value. By the end of last year he had invested $200,000 in these notes brought here by travellers. He then inclosed them in letters, and sent them to their proper destinations to be redeemed. Redeemed they were in due time, and the proceeds remitted in gold. In this business he earned the neat profit of $22,222, and the country was that much richer thereby. But Mr. Greeley, who only looked at the import of K's gold remittance, declared the country $22,222 worse off than before, and dares us to "come on" with the figures.

L, and some fifty thousand other skedaddlers ran off to Canada when the war broke out, for fear they might be drafted. Together with the colored folks who fled there, and the many travellers who went there from time to time, they carried with them most of our silver half-dollars, quarters, dimes, half-dimes, and three-cent pieces. These amounted to $25,000,000, which the skedaddlers, the colored folks, and the travellers, as with returning peace they slowly straggled back into the country, invested in Canadian knick-knacks, which they disposed of in the United States. The incoming goods were duly entered at our frontier custom-houses, but the outgoing silver was not. Mr. Greeley, unaware of this fact, detects an over-importation of $25,000,000, and is waiting to be elected to Congress in order to legislate the matter right.

M, (an actual transaction) had $1,000,000 in Illinois Central Railroad bonds, for which he desired to obtain $1,000,000 worth of iron rails to repair the road with. Not being able to effect the transaction in the

United States, he sent the bonds to Germany, where they were sold, and the proceeds invested in English railroad iron, worth $1,000,000 in Glasgow, but $1,100,000 in Chicago, ex duty, and plus transportation. By this transaction M, besides effecting the desired exchange, netted a profit of $100,000. Yet, according to the Commerce and Navigation Reports, and Mr. Greeley's one eye, as there had been no exports and $1,000,000 of imports, the country was a sufferer by the latter sum.

N, was a body of incorporators who owned a tract of land lying in the bend of a river. Standing in need of water power for manufacturing purposes, they resolved to cut a canal across the bend. As this would essentially benefit the navigation of the river, the State agreed to guaranty their bonds for a loan of money to the extent of $1,000,000. Finding no purchaser for these bonds in the United States, they remitted them to Europe, and there sold them at par. With the proceeds they purchased army blankets for the Boston market, on which they realized ten per cent. net profit. These sold, the avails were invested in barrows, spades, waterwheels, wages, &c., and in good time the canal was cut and the manufactory set a-going. Profitable as this thing was to N, Mr. Greeley's single-barrelled telescope sees in it only a loss to the country of $1,000,000.

O, represents the Illinois Central, Union Pacific, and other western railroads, owning grants of land along their respective roads, to sell which to actual settlers they open agencies in London, Havre, Antwerp, and other European cities. The emigrants who buy these lands pay for them in Europe, and set sail for America with their title-deeds in their pockets, and their axes on their shoulders, ready for a conquest over forest and prairie. The agents of the Illinois Central Railroad (see report of the Company), who have sold 1,664,422 acres, say at an average of ten dollars per acre, invested the proceeds, $16,644,220, in iron rails for the road, worth that sum in England, but ten per cent. more in Illinois, less duty and plus transportation. The road has thus not only netted a profit of $1,664,422 on the transaction, but sold their wild lands to actual settlers, who will soon convert them into productive farms. But Mr. Greeley, upon seeing an import of $16,644,220 of iron rails, declares the thing must be stopped or the country will perish.

P, is Sir Morton Peto and other European capitalists, who, believing that eight per cent., the average rate of interest in the United States, is better than three per cent., the average rate in England, invest $10,000,000 of capital in American enterprises. This capital is sent hither in the form of merchandise, to stock our railroads, farms, factories, etc., and is so much clear benefit to the country; but to Mr. Greeley's solitary vision it is only a curse.

Q, and his friends are cozy old-fashioned merchants in Boston city, who own one hundred and seventy-nine vessels (see Consular Reports, 1865), which trade between foreign ports and away from the United States altogether. These vessels have an aggregate burden of one million tons, are worth forty dollars, gold, per ton, and earn a net profit per annum of ten per cent. on their cost. Although in this kind of carrying trade we are wofully behind other nations, yet it yields, in twelve years (the average age of the vessels engaged in it), the neat little profit of $48,000,000, which is invested by Q in tea, coffee, and sugar, and imported into the United States at a net profit of ten per cent. Although an unquestionable gain to Q and the country at large of $52,800,000, Mr. Greeley, with his contracted views, only regards it as a dead loss on the import side of our Commerce and Navigation Returns.

R, was a bank which had a defaulting cashier, who ran away in 1857 with $500,000 of its funds. (Schuyler carried off a million of New Haven Railroad bonds). These funds were recovered and converted into gold, which was shipped to the United States. According to Mr. Greeley, who could find no record of exports to counterbalance it, the same was a dead loss to the country.

S, and his friends own 76,990 tons of whaling ships (see Commerce and Navigation Reports, 1866), worth $40 per ton, gold, or $3,079,600. These ships are sent annually to the Arctic regions and earn for S and his friends ten per cent., or $307,960 net profit each year. Five years' profits, consisting of whale oil, bone, etc., which, after an active and profitable trade at the Sandwich Islands, they returned with this year, were valued at $1,655,659, and were duly entered among the imports, furnishing to Mr. Greeley an indubitable proof that the country was losing money in this business, and that the attention of Congress should at once be directed toward supplying a proper remedy.

T, was a South American refugee, who brought with him a million of dollars in gold doubloons. After living here for many years, by which time, through foreign trading, his capital had doubled, he invested the entire avails in United States bonds, as a last and striking evidence of his faith in our institutions, and departed to his native country, there to rest his bones. This man clearly prospered, and so did the country in which he settled, and on whose national faith he lent all his fortune. Yet Mr. Greeley concludes the whole thing to have been a bad job for us, and harps upon another over-importation of $1,000,000.

U, is a gallant Yankee sea-captain, who picks up an abandoned vessel at sea laden with a valuable cargo of teas, and bravely tows her into port, receiving $200,000 of the proceeds of the sale of her cargo as salvage for his skill and intrepidity. From Mr. Greeley's point of view U is a traitor to his country, and suffering a merited poverty for over-importing. But U drives his carriage about town, and has his own opinion of Mr. Greeley's views.

V, having a debt of $300,000 due to him by a merchant in Alexandria, requests him to invest the same in Arabian horses, as fancy stock to improve American breeds. The horses arrive in good order, and on being sold, yield V a net profit of $30,000, besides enriching our native breeds of these useful animals. Mr. Greeley still holds out, and jots the whole transaction down as an additional evidence of national decadence.

W, X, Y, Z, represent 43,628,427,835,109 other commercial transactions, in all of which the parties to them and the countries in which they live make money, but which, regarded from Mr. Greeley's solitary point of view, should be stopped at once by appropriate legislation.

These various transactions, it will be perceived, have netted to the individuals engaged in them a clear profit of $66,391,813, while the country has added to its immediate stock of wealth not only this sum, but $58,344,220 over, viz: $124,736,033; while, according to the Balance of Trade chimera, which simply weighs the custom-house reports of the value of the exports with that of the imports (and their values in their respective countries of production, too), this commerce has been a loss to the country of $163,622,611—$11,000,000: $152,622,611.

So much for *theory* when confronted with *practice*.

The truth is, that the theory of the Balance of Trade should be precisely *reversed*. The profits accruing to the nation from any foreign commerce should be calculated by the overplus of the importation above the exportation. This overplus, after the deduction of expenses, is the real gain.

Here we have the true theory, and it is one which leads directly to freedom in trade. I now, gentlemen, abandon you this theory, as I have done all those of the preceding chapters. Do with it as you please, exaggerate it as you will; it has nothing to fear. Push it to the furthest extreme; imagine, if it so please you, that foreign nations should inundate us with useful produce of every description, and ask nothing in return; that our importations should be *infinite,* and our exportations *nothing.* Imagine all this, and still I defy you to prove that we will be the poorer in consequence.

Chapter 7

A PETITION

Petition from the Manufacturers of Candles, Wax-Lights, Lamps, Chandeliers, Reflectors, Snuffers, Extinguishers; and from the Producers of Tallow, Oil, Resin, Petroleum, Kerosene, Alcohol, and generally of every thing used for lights.

" *To the Honorable the Senators and Representatives of the United States in Congress assembled.*

"GENTLEMEN:—You are in the right way: you reject abstract theories; abundance, cheapness, concerns you little. You are entirely occupied with the interest of the producer, whom you are anxious to free from foreign competition. In a word, you wish to secure the *national market* to *national labor.*

"We come now to offer you an admirable opportunity for the application of your—what shall we say? your theory? no, nothing is more deceiving than theory—your doctrine? your system? your principle? But you do not like doctrines; you hold systems in horror; and, as for principles, you declare that there are no such things in political economy. We will say, then, your practice; your practice without theory, and without principle.

"We are subjected to the intolerable competition of a *foreign rival,* who enjoys, it would seem, such superior facilities for the production of light, that he is enabled to *inundate* our *national market* at so exceedingly reduced a price, that, the moment he makes his appearance, he draws off all custom from us; and thus an important branch of American industry, with all its innumerable ramifications, is suddenly reduced to a state of complete stagnation. This rival, who is no other than the sun, carries on so bitter a war against us, that we have every reason to believe that he has been excited to this course by our perfidious cousins, the Britishers. (Good diplomacy this, for the present time!) In this belief we are confirmed by the fact that in all his transactions with their befogged island, he is much more moderate and careful than with us.

"Our petition is, that it would please your Honorable Body to pass a law whereby shall be directed the shutting up of all windows, dormers, sky-lights, shutters, curtains—in a word, all openings, holes, chinks, and fissures through which the light of the sun is used to penetrate into our dwellings, to the prejudice of the profitable manufactures which we flatter ourselves we have been enabled to bestow upon the country; which country cannot, therefore, without ingratitude, leave us now to struggle unprotected through so unequal a contest.

"We pray your Honorable Body not to mistake our petition for a satire, nor to repulse us without at least hearing the reasons which we have to advance in its favor.

"And first, if, by shutting out as much as possible all access to natural light, you thus create the necessity for artificial light, is there in the United States an industrial pursuit which will not, through some connection with this important object, be benefited by it?

"If more tallow be consumed, there will arise a necessity for an increase of cattle and sheep. Thus artificial meadows must be in greater demand; and meat, wool, leather, and above all, manure, this basis of agricultural riches, must become more abundant.

"If more oil be consumed, it will effect a great impetus to our petroleum trade. Pit-Hole, Tack, and Oil Creek stock will go up exceedingly, and an immense revenue will thereby accrue to the numerous possessors of oil lands, who will be able to pay such a large tax that the national debt can be paid off at once. Besides that, the patent hermetical barrel trade, and numerous other industries connected with the oil trade, will prosper at an unprecedented rate, to the great benefit and glory of the country.

"Navigation would equally profit. Thousands of vessels would soon be employed in the whale fisheries, and thence would arise a navy capable of sustaining the honor of the United States, and of responding to the patriotic sentiments of the undersigned petitioners, candle-merchants, &c.

"But what words can express the magnificence which New York will then exhibit! Cast an eye upon the future, and behold the gildings, the bronzes, the magnificent crystal chandeliers, lamps, lusters, and candelabras, which will glitter in the spacious stores, compared to which the splendor of the present day will appear little and insignificant.

"There is none, not even the poor manufacturer of resin in the midst of his pine forests, nor the miserable miner in his dark dwelling, but who would enjoy an increase of salary and of comforts.

"Gentlemen, if you will be pleased to reflect, you cannot fail to be convinced that there is perhaps not one American, from the opulent stockholder of Pit-Hole, down to the poorest vender of matches, who is not interested in the success of our petition.

"We foresee your objections, gentlemen; but there is not one that you can oppose to us which you will not be obliged to gather from the works of the partisans of free trade. We dare challenge you to pronounce one word against our petition, which is not equally opposed to your own practice and the principle which guides your policy.

"If you tell us that, though we may gain by this protection, the United States will not gain, because the consumer must pay the price of it, we answer you:

"You have no longer any right to cite the interest of the consumer. For whenever this has been found to compete with that of the producer, you have invariably sacrificed the first. You have done this to *encourage labor,* to *increase the demand for labor.* The same reason should now induce you to act in the same manner.

"You have yourselves already answered the objection. When you were told: The consumer is interested in the free introduction of iron, coal, corn, wheat, cloths, &c., your answer was: Yes, but the producer is interested in their exclusion. Thus, also, if the consumer is interested in the admission of light, we, the producers, pray for its interdiction.

"You have also said the producer and the consumer are one. If the manufacturer gains by protection, he will cause the agriculturist to gain also; if agriculture prospers, it opens a market for manufactured goods. Thus we, if you confer upon us the monopoly of furnishing light during the day, will as a first consequence buy large quantities of tallow, coal, oil, resin, kerosene, wax, alcohol, silver, iron, bronze, crystal, for the supply of our business; and then we and our numerous contractors having become rich, our consumption will be great, and will become a means of contributing to the comfort and competency of the workers in every branch of national labor.

"Will you say that the light of the sun is a gratuitous gift, and that to repulse gratuitous gifts is to repulse riches under pretence of encouraging the means of obtaining them?

"Take care—you carry the death-blow to your own policy. Remember that hitherto you have always repulsed foreign produce, *because* it was an approach to a gratuitous gift, and *the more in proportion* as this approach was more close. You have, in obeying the wishes of other monopolists, acted only from a *half-motive;* to grant our petition there is a much *fuller inducement.* To repulse us, precisely for the reason that our case is a more complete one than any which have preceded it, would be to lay down the following equation: + x + = -; in other words, it would be to accumulate absurdity upon absurdity.

"Labor and Nature concur in different proportions, according to country and climate, in every article of production. The portion of Nature is always gratuitous; that of labor alone regulates the price.

"If a Lisbon orange can be sold at one hundredth the price of a New York one, it is because a natural and gratuitous heat does for the one, what the other only obtains from an artificial and consequently expensive one.

"When, therefore, we purchase a Portuguese orange, we may say that we obtain it 99/100 gratuitously and 1/100 by the right of labor; in other words, at a mere song compared to those of New York.

"Now it is precisely on account of this 99/100 *gratuity* (excuse the phrase) that you argue in favor of exclusion. How, you say, could national labor sustain the competition of foreign labor, when the first has every thing to do, and the last is rid of nearly all the trouble, the sun taking the rest of the business upon himself? If then the 99/100 *gratuity* can determine you to check competition, on what principle can the *entire gratuity* be alleged as a reason for admitting it? You are no logicians if, refusing the 99/100 gratuity as hurtful to human labor, you do not *a fortiori,* and with double zeal, reject the full gratuity.

"Again, when any article, as coal, iron, cheese, or cloth, comes to us from foreign countries with less labor than if we produced it ourselves, the difference in price is a gratuitous gift conferred upon us; and the gift is more or less considerable, according as the difference is greater or less. It is the quarter, the half, or the three-quarters of the value of the produce, in proportion as the foreign merchant requires the three-quarters, the half, or the quarter of the price. It is as complete as possible when the producer offers, as the sun does with light, the whole, in free gift. The question is, and we put it formally, whether you wish for the United States the benefit of gratuitous consumption, or the supposed advantages of laborious production. Choose: but be consistent. And does it not argue the greatest inconsistency to check, as you do, the importation of iron-ware, dry-goods, and other foreign manufactures, merely because, and even in proportion as, their price approaches zero, while at the same time you freely admit, and without limitation, the light of the sun, whose price is during the whole day at zero?"

Chapter 8

DISCRIMINATING DUTIES

A poor laborer of Ohio had raised, with the greatest possible care and attention, a nursery of vines, from which, after much labor, he at last succeeded in producing a pipe of Catawba wine, and forgot, in the joy of his success, that each drop of this precious nectar had cost a drop of sweat to his brow.

"I will sell it," said he to his wife, "and with the proceeds I will buy lace, which will serve you to make a present for our daughter."

The honest countryman, arriving in the city of Cincinnati, there met an Englishman and a Yankee.

The Yankee said to him, "Give me your wine, and I in exchange will give you fifteen bundles of Yankee lace."

The Englishman said, "Give it to me, and I will give you twenty bundles of English lace, for we English can spin cheaper than the Yankees."

But a custom-house officer standing by, said to the laborer, "My good fellow, make your exchange, if you choose, with Brother Jonathan, but it is my duty to prevent your doing so with the Englishman."

"What!" exclaimed the countryman, "you wish me to take fifteen bundles of New England lace, when I can have twenty from Manchester!"

"Certainly," replied the custom-house officer; "do you not see that the United States would be a loser if you were to receive twenty bundles instead of fifteen?"

"I can scarcely understand this," said the laborer.

"Nor can I explain it," said the custom-house officer, "but there is no doubt of the fact; for congressmen, ministers, and editors, all agree that a people is impoverished in proportion as it receives a large compensation for any given quantity of its produce."

The countryman was obliged to conclude his bargain with the Yankee. His daughter received but three-fourths of her present; and these good folks are still puzzling themselves to discover how it can happen that people are ruined by receiving four instead of three; and why they are richer with three dozen bundles of lace instead of four.

Chapter 9

A WONDERFUL DISCOVERY

At this moment, when all minds are occupied in endeavoring to discover the most economical means of transportation; when, to put these means into practice, we are levelling roads, improving rivers, perfecting steamboats, establishing railroads, and attempting various systems of traction, atmospheric, hydraulic, pneumatic, electric, &c.; at this moment, when, I believe, every one is seeking in sincerity and with ardor the solution of this problem—"*To bring the price of things in their place of consumption, as near as possible to their price in that of production*"—I would believe myself to be acting a culpable part towards my country, towards the age in which I live, and towards myself, if I were longer to keep secret the wonderful discovery which I have just made.

I am well aware that the self-illusions of inventors have become proverbial, but I have, nevertheless, the most complete certainty of having discovered an infallible means of bringing produce from all parts of the world into the United States, and reciprocally to transport ours, with a very important reduction of price.

Infallible! and yet this is but a single one of the advantages of my astonishing invention, which requires neither plans nor devices, neither preparatory studies, nor engineers, nor machinists, nor capital, nor stockholders, nor governmental assistance! There is no danger of shipwrecks, of explosions, of shocks of fire, nor of displacement of rails! It can be put into practice without preparation almost any day we think proper!

Finally: and this will, no doubt, recommend it to the public, it will not increase the Budget one cent; but the contrary. It will not augment the number of office-holders, nor the exigencies of State; but the contrary. It will put in hazard the liberty of no one; but on the contrary, it will secure to each a greater freedom.

I have been led to this discovery, not from accident, but from observation, and I will tell you how.

I had this question to determine:

"Why does any article made, for instance, at Montreal, bear an increased price on its arrival at New York?"

It was immediately evident to me that this was the result of *obstacles* of various kinds existing between Montreal and New York. First, there is *distance,* which cannot be overcome without trouble and loss of time; and either we must submit to these troubles and losses in our own person, or pay another for bearing them for us. Then come rivers, hills, accidents, heavy and muddy roads. These are so many *difficulties* to be overcome; in order to do which, causeways are constructed, bridges built, roads cut and paved, railroads established, &c. But all this is costly, and the article transported must bear its portion of the expense. There are robbers, too, on the roads, sometimes, and this necessitates railway guards, a police force, &c.

Now, among these *obstacles,* there is one which we ourselves have lately placed, and that at no little expense, between Montreal and New York. This consists of men planted along the frontier, armed to the teeth, whose business it is to place *difficulties* in the way of the transportation of goods from one country to another. These men are called custom-house officers, and their effect is precisely similar to that of rutted and boggy roads. They retard and put obstacles in the way of transportation, thus contributing to the difference which we have remarked between the price of production and that of consumption; to diminish which difference, as much as possible, is the problem which we are seeking to resolve.

Here, then, we have found its solution. Let our tariff be diminished: we will thus have constructed a Northern railway which will cost us nothing. Nay, more, we will be saved great expenses, and will begin, from the first day, to save capital.

Really, I cannot but ask myself, in surprise, how our brains could have admitted so whimsical a piece of folly as to induce us to pay many millions to destroy the *natural obstacles* interposed between the United States and other nations, only at the same time to pay so many millions more in order to replace them by *artificial obstacles,* which have exactly the same effect; so that the obstacle removed and the obstacle created, neutralize each other, things go on as before, and the only result of our trouble is a double expense.

An article of Canadian production is worth, at Montreal, twenty dollars, and, from the expenses of transportation, thirty dollars at New York. A similar article of New York manufacture costs forty dollars. What is our course under these circumstances?

First, we impose a duty of at least ten dollars on the Canadian article, so as to raise its price to a level with that of the New York one—the government, withal, paying numerous officials to attend to the levying of this duty. The article thus pays ten dollars for transportation, and ten for the tax.

This done, we say to ourselves: Transportation between Montreal and New York is very dear; let us spend two or three millions in railways, and we will reduce it one-half. Evidently the result of such a course will be to get the Canadian article at New York for thirty-five dollars, viz.:

20 dollars—price at Montreal. 10 " duty. 5 " transportation by railway.—35 dollars—total, or market price at New York.

Could we not have attained the same end by lowering the tariff to five dollars? We would then have—

20 dollars—price at Montreal. 5 " duty. 10 " transportation on the common road.—35 dollars—total, or market price at New York.

And this arrangement would have saved us the $2,000,000 spent upon the railway, besides the expense saved in custom-house surveillance, which would of course diminish in proportion as the temptation to smuggling would become less.

But it is answered: The duty is necessary to protect New York industry. So be it; but do not then destroy the effect of it by your railway. For if you persist in your determination to keep the Canadian article on a par with the New York one at forty dollars, you must raise the duty to fifteen dollars, in order to have:—

20 dollars—price at Montreal.

15 protective duty.

5 transportation by railway.

40 dollars—total, at equalized prices.

And I now ask, of what benefit, under these circumstances, is the railway?

Frankly, is it not humiliating to the nineteenth century, that it should be destined to transmit to future ages the example of such puerilities seriously and gravely practised? To be the dupe of another, is bad enough; but to employ all the forms and ceremonies of representation in order to cheat oneself—to doubly cheat oneself, and that too in a mere numerical account—truly this is calculated to lower a little the pride of this *enlightened age.*

Chapter 10

RECIPROCITY

We have just seen that all which renders transportation difficult, acts in the same manner as protection; or, if the expression be preferred, that protection tends towards the same result as all obstacles to transportation.

A tariff may be truly spoken of as a swamp, a rut, a steep hill; in a word, an *obstacle,* whose effect is to augment the difference between the price of consumption and that of production. It is equally incontestable that a swamp, a bog, &c., are veritable protective tariffs.

There are people (few in number, it is true, but such there are) who begin to understand that obstacles are not the less obstacles because they are artificially created, and that our well-being is more advanced by freedom of trade than by protection; precisely as a canal is more desirable than a sandy, hilly, and difficult road.

But they still say, this liberty ought to be reciprocal. If we take off our taxes in favor of Canada, while Canada does not do the same towards us, it is evident that we are duped. Let us, then, make *treaties of commerce* upon the basis of a just reciprocity; let us yield where we are yielded to; let us make the *sacrifice* of buying that we may obtain the advantage of selling.

Persons who reason thus, are (I am sorry to say), whether they know it or not, governed by the protectionist principle. They are only a little more inconsistent than the pure protectionists, as these are more inconsistent than the absolute prohibitionists.

I will illustrate this by a fable:

There were, it matters not where, two towns, New York and Montreal, which, at great expense, had a road built, which connected them with each other. Some time after this was done, the inhabitants of New York became uneasy, and said: "Montreal is overwhelming us with its productions; this must be attended to." They established, therefore, a corps of *Obstructors,* so called, because their business was to place obstacles in the way of the convoys which arrived from Montreal. Soon after, Montreal also established a corps of Obstructors.

After some years, people having become more enlightened, the inhabitants of Montreal began to discover that these reciprocal obstacles might possibly be reciprocal injuries. They sent, therefore, an ambassador to New York, who (passing over the official phraseology) spoke much to this effect: "We have built a road, and now we put obstacles in the way of this road. This is absurd. It would have been far better to have left things in their original position, for then we would not have been put to the expense of building our road, and afterwards of creating difficulties. In the name of Montreal I come to propose to you not to renounce at once our system of mutual obstacles, for this would be acting according to a principle, and we despise principles as much as you do; but to somewhat lighten these obstacles, weighing at the same time carefully our respective *sacrifices.*" The ambassador having thus spoken, the town of New York asked time to reflect; manufacturers, office-seekers, congressmen, and custom-house officers, were consulted; and at last, after some years' deliberation, it was declared that the negotiations were broken off.

At this news, the inhabitants of Montreal held a council. An old man (who it has always been supposed had been secretly bribed by New York) rose and said: "The obstacles raised by New York are injurious to our sales; this is a misfortune. Those which we ourselves create, injure our purchases; this is a second misfortune. We have no power over the first, but the second is entirely dependent upon ourselves. Let us then at least get rid of one, since we cannot be delivered from both. Let us suppress our corps of Obstructors, without waiting for New York to do the same. Some day or other she will learn to better calculate her own interests."

A second counsellor, a man of practice and of facts, uncontrolled by principles and wise in ancestral experience, replied: "We must not listen to this dreamer, this theorist, this innovator, this Utopian, this political economist, this friend to New York. We would be entirely ruined if the embarrassments of the road were not carefully weighed and exactly equalized between New York and Montreal. There would be more difficulty in going than in coming; in exportation than in importation. We would be with regard to New York, in the inferior condition in which Havre, Nantes, Bordeaux, Lisbon, London, Hamburg, and New Orleans, are, in relation to cities placed higher up the rivers Seine, Loire, Garonne, Tagus, Thames, Elbe, and Mississippi; for the difficulties of ascending must always be greater than those of descending rivers."

"(A voice exclaims: 'But the cities near the mouths of rivers have always prospered more than those higher up the stream.')

"This is not possible."

"(The same voice: 'But it is a fact.')

"Well, they have then prospered *contrary to rule.*"

Such conclusive reasoning staggered the assembly. The orator went on to convince them thoroughly and conclusively by speaking of national independence, national honor, national dignity, national labor, overwhelming importation, tributes, ruinous competition. In short, he succeeded in determining the assembly to continue their system of obstacles, and I can now point out a certain country where you may see road-workers and Obstructors working with the best possible understanding, by the decree of the same legislative assembly, paid by the same citizens; the first to improve the road, the last to embarrass it.

Chapter 11

ABSOLUTE PRICES

If we wish to judge between freedom of trade and protection, to calculate the probable effect of any political phenomenon, we should notice how far its influence tends to the production of *abundance* or *scarcity*, and not simply of *cheapness* or *dearness* of price. We must beware of trusting to absolute prices: it would lead to inextricable confusion.

Mr. Protectionist, after having established the fact that protection raises prices, adds:

"The augmentation of price increases the expenses of life, and consequently the price of labor, and every one finds in the increase of the price of his produce the same proportion as in the increase of his expenses. Thus, if everybody pays as consumer, everybody receives also as producer."

It is evident that it would be easy to reverse the argument, and say: If everybody receives as producer, everybody must pay as consumer.

Now what does this prove? Nothing whatever, unless it be that protection *transfers* riches, uselessly and unjustly. Spoliation does the same.

Again, to prove that the complicated arrangements of this system give even simple compensation, it is necessary to adhere to the "*consequently*" of Mr. Protectionist, and to convince oneself that the price of labor rises with that of the articles protected. This is a question of fact. For my own part I do not believe in it, because I think that the price of labor, like everything else, is governed by the proportion existing between the supply and the demand. Now I can perfectly well understand that *restriction* will diminish the supply of produce, and consequently raise its price; but I do not as clearly see that it increases the demand for labor, thereby raising the rate of wages. This is the less conceivable to me, because the sum of labor required depends upon the quantity of disposable capital; and protection, while it may change the direction of capital, and transfer it from one business to another, cannot increase it one penny.

This question, which is of the highest interest, we will examine elsewhere. I return to the discussion of *absolute prices,* and declare that there is no absurdity which cannot be rendered specious by such reasoning as that which is commonly resorted to by protectionists.

Imagine an isolated nation possessing a given quantity of cash, and every year wantonly burning the half of its produce; I will undertake to prove by the protective theory that this nation will not be the less rich in consequence of such a procedure. For, the result of the conflagration must be, that everything would double in price. An inventory made before this event, would offer exactly the same nominal value as one made after it. Who, then, would be the loser? If John buys his cloth dearer, he also sells his corn at a higher price; and if Peter makes a loss on the purchase of his corn, he gains it back by the sale of his cloth. Thus "every one finds in the increase of the price of his produce, the same proportion as in the increase of his expenses: and thus if everybody pays as consumer, everybody also receives as producer."

All this is nonsense, and not science.

The simple truth is, that whether men destroy their corn and cloth by fire, or by use, the effect is the same as regards price, but not as regards riches, for it is precisely in the enjoyment of the use, that riches—in other words, comfort, well-being—exist.

Restriction may in the same way, while it lessens the abundance of things, raise their prices, so as to leave each individual as rich, *numerically speaking,* as when unembarrassed by it. But because we put down in an inventory three bushels of corn at $1, or four bushels at 75 cents, and sum up the nominal value of each inventory at $3, does it thence follow that they are equally capable of contributing to the necessities of the community?

To this truthful and common-sense view of the phenomenon of consumption it will be my continual endeavor to lead the protectionists; for in this is the end of all my efforts, the solution of every problem. I must continually repeat to them that restriction, by impeding commerce, by limiting the division of labor, by forcing it to combat difficulties of situation and temperature, must in its results diminish the quantity produced by any fixed quantum of labor. And what can it benefit us that the smaller quantity produced under the protective system bears the same *nominal value* as the greater quantity produced under the free trade system? Man does not live on *nominal values,* but on real articles of produce; and the more abundant these articles are, no matter what price they may bear, the richer is he.

The following passage occurs in the writings of a French protectionist:

"If fifteen millions of merchandise sold to foreign nations, be taken from our ordinary produce, calculated at fifty millions, the thirty-five millions of merchandise which remain, not being sufficient for the ordinary demand, will increase in price to the value of fifty millions. The revenue of the country will thus represent fifteen millions more in value. . . . There will then be an increase of fifteen millions in the riches of the country; precisely the amount of the importation of money."

To sum up our judgment of the two systems, let us contemplate their different effects when pushed to the most exaggerated extreme. This is droll enough! If a country has made in the course of the year fifty millions of revenue in harvests and merchandise, she need but sell one-quarter to foreign nations, in order to make herself one-quarter richer than before! If then she sold the half, she would increase her riches by one-half; and if the last hair of her wool, the last grain of her wheat, were to be changed for cash, she would thus raise her product to one hundred millions, where before it was but fifty! A singular manner, certainly, of becoming rich. Unlimited price produced by unlimited scarcity!

According to the protectionist just quoted, the French would be quite as rich, that is to say, as well provided with everything, if they had but a thousandth part of their annual produce, because this part would then be worth a thousand times its natural value! So much for looking at prices alone.

According to us, the French would be infinitely rich if their annual produce were infinitely abundant, and consequently bearing no value at all.

Chapter 12

DOES PROTECTION RAISE THE RATE OF WAGES?

When we hear our beardless scribblers, romancers, reformers, our perfumed magazine writers, stuffed with ices and champagne, as they carefully place in their portfolios the sentimental scissorings which fill the current literature of the day, or cause to be decorated with gilded ornaments their tirades against the egotism and the individualism of the age; when we hear them declaiming against social abuses, and groaning over deficient wages and needy families; when we see them raising their eyes to heaven and weeping over the wretchedness of the laboring classes, while they never visit this wretchedness unless it be to draw lucrative sketches of its scenes of misery, we are tempted to say to them: The sight of you is enough to make me sicken of attempting to teach the truth.

Affectation! Affectation! It is the nauseating disease of the day! If a thinking man, a sincere philanthropist, takes into consideration the condition of the working classes and endeavors to lay bare their necessities, scarcely has his work made an impression before it is greedily seized upon by the crowd of reformers, who turn, twist, examine, quote, exaggerate it, until it becomes ridiculous; and then, as sole compensation, you are overwhelmed with such big words as: Organization, Association; you are flattered and fawned upon until you become ashamed of publicly defending the cause of the working man; for how can it be possible to introduce sensible ideas in the midst of these sickening affectations?

But we must put aside this cowardly indifference, which the affectation that provokes it is not enough to justify.

Working men, your situation is singular! You are robbed, as I will presently prove to you. But no: I retract the word; we must avoid an expression which is violent; perhaps, indeed, incorrect; inasmuch as this spoliation, wrapped in the sophisms which disguise it, is practised, we must believe, without the intention of the spoiler, and with the consent of the spoiled. But it is nevertheless true that you are deprived of the just remuneration of your labor, while no one thinks of causing *justice* to be rendered to you. If you could be consoled by the noisy appeals of your champions to philanthropy, to powerless charity, to degrading almsgiving, or if the high-sounding words of Voice of the People, Rights of Labor, &c., would relieve you—these indeed you can have in abundance. But *justice,* simple *justice*—this nobody thinks of rendering you. For would it not be *just* that after a long day's labor, when you have received your wages, you should be permitted to exchange them for the largest possible sum of comforts you can obtain voluntarily from any man upon the face of the earth?

I too, perhaps, may some day speak to you of the Voice of the People, the Rights of Labor, &c., and may perhaps be able to show you what you have to expect from the chimeras by which you allow yourselves to be led astray.

In the meantime let us examine if *injustice* is not done to you by the legislative limitation of the number of persons from whom you are allowed to buy those things which you need; as iron, coal, cotton and woollen cloths, &c.; thus artificially fixing (so to express myself) the price which these articles must bear.

Is it true that protection, which avowedly raises prices, and thus injures you, proportionably raises the rate of wages?

On what does the rate of wages depend?

One of your own class has energetically said: "When two workmen run after a boss, wages fall; when two bosses run after a workman, wages rise."

Allow me, in similar laconic phrase, to employ a more scientific, though perhaps a less striking expression: "The rate of wages depends upon the proportion which the supply of labor bears to the demand."

On what depends the *demand* for labor?

On the quantity of disposable capital seeking investment. And the law which says, "Such or such an article shall be limited to home production and no longer imported from foreign countries," can it in any degree increase this capital? Not in the least. This law may withdraw it from one course, and transfer it to another; but cannot increase it one penny. Then it cannot increase the demand for labor.

While we point with pride to some prosperous manufacture, can we answer, whence comes the capital with which it is founded and maintained? Has it fallen from the moon? or rather is it not drawn either from agriculture, or stock-breeding, or commerce? We here see why, since the reign of protective tariffs, if we see more workmen in our mines and our manufacturing towns, we find also fewer vessels in our ports, fewer graziers and fewer laborers in our fields and upon our hill-sides.

I could speak at great length upon this subject, but prefer illustrating my thought by an example.

A countryman had twenty acres of land, with a capital of $10,000. He divided his land into four parts, and adopted for it the following changes of crops: 1st, maize; 2d, wheat; 3d, clover; and 4th, rye. As he needed for himself and family but a small portion of the grain, meat, and dairy produce of the farm, he sold the surplus and bought iron, coal, cloths, etc. The whole of his capital was yearly distributed in wages and payments of accounts to the workingmen of the neighborhood. This capital was, from his sales, again returned to him, and even increased from year to year. Our countryman, being fully convinced that idle capital produces nothing, caused to circulate among the working classes this annual increase, which he devoted to the inclosing and clearing of lands, or to improvements in his farming utensils and his buildings. He deposited some sums in reserve in the hands of a neighboring banker, who on his part did not leave these idle in his strong-box, but lent them to various tradesmen, so that the whole came to be usefully employed in the payment of wages.

The countryman died, and his son, become master of the inheritance, said to himself: "It must be confessed that my father has, all his life, allowed himself to be duped. He bought iron, and thus paid *tribute* to England, while our own land could, by an effort, be made to produce iron as well as England. He bought coal, cloths, and oranges, thus paying *tribute* to New Brunswick, France, and Sicily, very unnecessarily; for coal may be found, doeskins may be made, and oranges may be forced to grow, within our own territory. He paid tribute to the foreign miner and the weaver;

our own servants could very well mine our iron and get up native doe-skins almost as good as the French article. He did all he could to ruin himself, and gave to strangers what ought to have been kept for the benefit of his own household."

Full of this reasoning, our headstrong fellow determined to change the routine of his crops. He divided his farm into twenty parts. On one he dug for coal; on another he erected a cloth factory; on a third he put a hot-house and cultivated the orange; he devoted the fourth to vines, the fifth to wheat, &c., &c. Thus he succeeded in rendering himself *independent,* and furnished all his family supplies from his own farm. He no longer received anything from the general circulation; neither, it is true, did he cast anything into it. Was he the richer for this course? No; for his mine did not yield coal as cheaply as he could buy it in the market, nor was the climate favorable to the orange. In short, the family supply of these articles was very inferior to what it had been during the time when the father had obtained them and others by exchange of produce.

With regard to the demand for labor, it certainly was no greater than formerly. *There were, to be sure, five times as many fields to cultivate, but they were five times smaller.* If coal was mined, there was also less wheat; and because there were no more oranges bought, neither was there any more rye sold. Besides, the farmer could not spend in wages more than his capital, and his capital, instead of increasing, was now constantly diminishing. A great part of it was necessarily devoted to numerous buildings and utensils, indispensable to a person who determines to undertake everything. In short, the supply of labor continued the same, but the means of paying became less.

The result is precisely similar when a nation isolates itself by the prohibitive system. Its number of industrial pursuits is certainly multiplied, but their importance is diminished. In proportion to their number, they become less productive, for the same capital and the same skill are obliged to meet a greater number of difficulties. The fixed capital absorbs a greater part of the circulating capital; that is to say, a greater part of the funds destined to the payment of wages. What remains, ramifies itself in vain; the quantity cannot be augmented. It is like the water of a deep pond, which, distributed among a multitude of small reservoirs, appears to be more abundant, because it covers a greater quantity of soil, and presents a larger surface to the sun, while we hardly perceive that, precisely on this account, it absorbs, evaporates, and loses itself the quicker.

Capital and labor being given, the result is, a sum of production, always the less great in proportion as obstacles are numerous. There can be no doubt that international barriers, by forcing capital and labor to struggle against greater difficulties of soil and climate, must cause the general production to be less, or, in other words, diminish the portion of comforts which would thence result to mankind. If, then, there be a general diminution of comforts, how, working men, can it be possible that *your* portion should be increased? Under such a supposition it would be necessary to believe that the rich, those who made the law, have so arranged matters, that not only they subject themselves to their own proportion of the general diminution, but taking the whole of it upon themselves, that they submit also to a further loss in order to increase your gains. Is this credible? Is this possible? It is, indeed, a most suspicious act of generosity; and if you act wisely you will reject it.

Chapter 13

THEORY AND PRACTICE

Defenders of free trade, we are accused of being mere theorists, of not giving sufficient weight to the practical.

"What a fearful charge against you, free traders," say the protectionists, "is this long succession of distinguished statesmen, this imposing race of writers, who have all held opinions differing from yours!" This we do not deny. We answer, "It is said, in support of established errors, that 'there must be some foundation for ideas so generally adopted by all nations. Should not one distrust opinions and arguments which overturn that which, until now, has been held as settled; that which is held as certain by so many persons whose intelligence and motives make them trustworthy?'"

We confess this argument should make a profound impression, and ought to throw doubt on the most incontestable points, if we had not seen, one after another, opinions the most false, now generally acknowledged to be such, received and professed by all the world during a long succession of centuries. It is not very long since all nations, from the most rude to the most enlightened, and all men, from the street-porter to the most learned philosopher, believed in the four elements. Nobody had thought of contesting this doctrine, which is, however, false; so much so, that at this day any mere naturalist's assistant, who should consider earth, water, and fire, elements, would disgrace himself.

On which our opponents make this observation: "If you suppose you have thus answered the very forcible objection you have proposed to yourselves, you deceive yourselves strangely. Suppose that men, otherwise intelligent, should be mistaken on any point whatever of natural history for many centuries, that would signify or prove nothing. Would water, air, earth, fire, be less useful to man whether they were or were not elements? Such errors are of no consequence; they lead to no revolutions, do not unsettle the mind; above all, they injure no interests, so they might, without inconvenience, endure for millions of years. The physical world would progress just as if they did not exist. Would it be thus with errors which attack the moral world? Can we conceive that a system of government, absolutely false, consequently injurious, could be carried out through many centuries, among many nations, with the general consent of educated men? Can we explain how such a system could be reconciled with the ever-increasing prosperity of nations? You acknowledge that the argument you combat ought to make a profound impression. Yes, truly, and this impression remains, for you have rather strengthened than destroyed it."

Or again, they say: "It was only in the middle of the last century, the eighteenth century, in which all subjects, all principles, without exception, were delivered up to public discussion, that these furnishers of speculative ideas which are applied to everything without being applicable to anything—commenced writing on political economy. There existed, however, a system of political economy, not written, but practised by governments. It is said that Colbert was its inventor, and it was the rule of all the States of Europe. What is more singular, it has remained so till lately, despite anathemas and contempt, and despite the discoveries of the modern school. This system, which our writers have called the *mercantile system,* consists in opposing, by prohibitions and duties, such foreign productions as might ruin our manufacturers by their competition. This system has been pronounced futile, absurd, capable of ruining any country, by economical writers of all schools. It has been banished from all books, reduced to take refuge in the practice of every people; and we do not understand why, in regard to the wealth of nations, governments should not have yielded themselves to wise authors rather than to *the old experience* of a system. Above all, we cannot conceive why, in political economy, the American government should persist in resisting the progress of light, and in preserving, in its practice, those old errors which all our economists of the pen have designated. But we have said too much about this mercantile system, which has in its favor *facts* alone, though sustained by scarcely a single writer of the day."

Would not one say, who listened only to this language, that we political economists, in merely claiming for every one *the free disposition of his own property,* had, like the Fourierists, conjured up from our brains a new social order, chimerical and strange; a sort of phalanstery, without precedent in the annals of the human race, instead of merely talking plain *meum* and *tuum* It seems to us that if there is in all this anything utopian, anything problematical, it is not free trade, but protection; it is not the right to exchange, but tariff after tariff applied to overturning the natural order of commerce.

But it is not the point to compare and judge of these two systems by the light of reason; the question for the moment is, to know which of the two is founded upon experience.

So, Messrs. Monopolists, you pretend that the facts are on your side; that we have, on our side, theories only.

You even flatter yourselves that this long series of public acts, this old experience of the world, which you invoke, has appeared imposing to us, and that we confess we have not as yet refuted you as fully as we might.

But we do not cede to you the domain of facts, for you have on your side only exceptional and contracted facts, while we have universal ones to oppose to them; the free and voluntary acts of all men.

What do you say, and what say we?

We say:

"It is better to buy from others anything which would cost more to make ourselves."

And on your part you say:

"It is better to make things ourselves, even though it would cost less to purchase them from others."

Now, gentlemen, laying aside theory, demonstration, argument, everything which appears to afflict you with nausea, which of these assertions has in its favor the sanction of *universal practice?*

Visit the fields, work-rooms, manufactories, shops; look above, beneath, and around you; investigate what is going on in your own establishment; observe your own conduct at all times, and then say which is the principle that directs these labors, these workmen, these inventors, these merchants; say, too, which is your own individual practice.

Does the farmer make his clothes? Does the tailor raise the wheat which he consumes? Does not your housekeeper cease making bread at home so soon as she finds it more economical to buy it from the baker? Do you give up the pen for the brush in order to avoid paying tribute to the shoe-black? Does not the whole economy of society depend on the separation of occupations, on the division of labor; in one word, on *exchange?* And is exchange anything else than the calculation which leads us to discontinue, as far as we can, direct production, when indirect acquisition spares us time and trouble?

You are not, then, men of *practice,* since you cannot show a single man on the surface of the globe who acts in accordance with your principle.

"But," you will say, "we have never heard our principle made the rule of individual relations. We comprehend perfectly that this would break the social bond, and force men to live, like snails, each one in his own shell. We limit ourselves to asserting that it governs *in fact* the relations which are established among the agglomerations of the human family."

But still, this assertion is erroneous. The family, the village, the town, the county, the state, are so many agglomerations, which all, without any exception, *practically* reject your principle, and have never even thought of it. All of them procure, by means of exchange, that which would cost them more to procure by means of production. Nations would act in the same natural manner, if you did not prevent it *by force.*

As for you, you form a theory, in the unfavorable sense of the word. You imagine, you invent—proceedings which are not sanctioned by the practice of any living man under the vault of heaven—and then you call to your assistance constraint and prohibition. You need, indeed, have recourse to *force,* since, in wishing that men should *produce* that which it would be more advantageous to them to *buy,* you wish them to renounce an *advantage;* you demand that they should act in accordance with a doctrine which implies contradiction even in its terms.It is *we,* then, who are the men of practice and of experience; for, in order to combat the interdict which you have placed exceptionally on certain international exchanges, we appeal to the practice and experience of all individuals, and all agglomerations of individuals whose acts are voluntary, and consequently may be called on for testimony. But you commence by *constraining*, by *preventing*, and then you avail yourself of acts caused by prohibition to exclaim, "See! practice justifies us!" You oppose our *theory*, indeed all *theory*. But when you put a principle in antagonism with ours, do you, by chance, fancy that you have formed no *theory?* No, no; erase that from your plea. You form a theory as well as ourselves; but between yours and ours there is this difference: our theory consists merely in observing universal facts, universal sentiments, universal calculations and proceedings, and further, in classifying them and arranging them, in order to understand them better. It is so little opposed to practice, that it is nothing but *practice explained*. We observe the actions of men moved by the instinct of preservation and of progress; and what they do freely, voluntarily, is precisely what we call *political economy*, or the economy of society. We go on repeating with out cessation: "Every man is *practically* an excellent economist, producing or exchanging, according as it is most advantageous to him to exchange or to produce. Each one, through experience, is educated to science; or rather, science is only that same experience scrupulously observed and methodically set forth."

Now, this doctrine, which, you argue, would be absurd in individual relations, we defy you to extend, even in speculation, to transactions between families, towns, counties, states. By your own avowal, it is applicable to international relations only.

And this is why you are obliged to repeat daily: "Principles are not in their nature absolute. That which is *well* in the individual, the family, the county, the state, is *evil* in the nation. That which is *good* in detail—such as, to purchase rather than to produce, when purchase is more advantageous than production—is bad in the mass. The political economy of individuals is not that of nations," and other rubbish, *ejusdem farinae*. And why all this? Look at it closely. It is in order to prove to us that we, consumers, are your property, that we belong to you body and soul, that you have an exclusive right to our stomachs and limbs, and it is for you to nourish us and clothe us at your own price, however great may be your ignorance, your rapacity, or the inferiority of your position.

No, you are not men of practice; you are men of abstraction—and of extraction!

Chapter 14

CONFLICT OF PRINCIPLES

There is one thing which confounds us, and it is this:

Some sincere publicists, studying social economy from the point of view of producers only, have arrived at this double formula:

"Governments ought to dispose of the consumers subject to the influence of their laws, in favor of national labor."

"They should render distant consumers subject to their laws, in order to dispose of them in favor of national labor."

The first of these formulas is termed *protection;* the latter, *expediency.*

Both rest on the principle called Balance of Trade; the formula of which is:

"A people impoverishes itself when it imports, and enriches itself when it exports."

Of course, if every foreign purchase is a tribute paid, a loss, it is perfectly evident we must restrain, even prohibit, importations.

And if all foreign sales are tribute received, profit, it is quite natural to create channels of outlet, even by force.

Protective System—Colonial System: two aspects of the same theory. To *hinder* our fellow-citizens purchasing of foreigners, *to force* foreigners to purchase from our fellow-citizens, are merely two consequences of one identical principle. Now, it is impossible not to recognize that according to this doctrine, general utility rests on *monopoly,* or interior spoliation, and on *conquest,* or exterior spoliation.

Let us enter one of the cabins among the Adirondacks. The father of the family has received for his work only a slender salary. The icy northern blast makes his half naked children shiver, the fire is extinguished, and the table bare. There are wool, and wood, and coal, just over the St. Lawrence; but these commodities are forbidden to the family of the poor day-laborer, for the other side of the river is no longer the United States. The foreign pine-logs may not gladden the hearth of his cabin; his children may not know the taste of Canadian bread, the wool of Upper Canada will not bring back warmth to their benumbed limbs. General utility wills it so. All very well! but acknowledge that here it contradicts justice. To dispose by legislation of consumers, to limit them to the products of national labor, is to encroach upon their liberty, to forbid them a resource (exchange) in which there is nothing contrary to morality; in one word, it is to do them injustice.

"Yet this is necessary," it is said, "under the penalty of seeing national labor stopped, under the penalty of striking a fatal blow at public prosperity."

The writers of the protectionist school arrive then at this sad conclusion; that there is a radical incompatibility between justice and utility.

On the other side, if nations are interested in selling, and not in buying, violent action and reaction are the natural condition of their relations, for each will seek to impose its products on all, and all will do their utmost endeavor to reject the products of each.

As a sale, in effect, implies a purchase, and since, according to this doctrine, to sell is to benefit, as to buy is to injure, every international transaction implies the amelioration of one people, and the deterioration of another.

But, on one side, men are fatally impelled towards that which profits them: on the contrary, they resist instinctively whatever injures them; whence we must conclude that every people bears within itself a natural force of expansion, and a not less natural power of resistance, which are equally prejudicial to all the others; or, in other terms, that antagonism and war are the natural constitution of human society!

So that the theory which we are discussing may be summed up in these two axioms:

"Utility is incompatible with justice at home,"

"Utility is incompatible with peace abroad."

Now that which astonishes us, which confounds us, is, that a publicist, a statesman, who has sincerely adhered to an economic doctrine whose principle clashes so violently with other incontestable principles, could enjoy one moment's calm and repose of mind. As for us, it seems to us, that if we had penetrated into science by this entrance, if we did not clearly perceive that liberty, utility, justice, peace, are things not only compatible, but closely allied together, so to say, identical with each other, we would try to forget all we had learned; we would say to ourselves:

"How could God will that men shall attain prosperity only through injustice and war? How could He will that they may remove war and injustice only by renouncing their own well-being?"

Does not the science which has conducted us to the horrible blasphemy which this alternative implies deceive us by false lights; and shall we dare take on ourselves to make it the basis of legislation for a great people? And when a long succession of illustrious philosophers have brought together more comforting results from this same science, to which they have consecrated their whole lives; when they affirm that Liberty and Utility are reconciled with Justice and Peace, that all these grand principles follow infinite parallels, without clashing, throughout all eternity; have they not in their favor the presumption which results from all we know of the goodness and the wisdom of God, manifested in the sublime harmony of the material creation? Ought we lightly to believe, against such a presumption, and in face of so many imposing authorities, that it has pleased this same God to introduce antagonism and a discord into the laws of the moral world?

No, no; before taking it for granted that all social principles clash, shock, and neutralize each other, and are in anarchical, eternal, irremediable, conflict together; before imposing on our fellow citizens the impious system to which such reasoning conducts us, we had better go over the whole chain, and assure ourselves that there is no point on the way where we may have gone astray.

And if, after a faithful examination, twenty times recommenced, we should always return to this frightful conclusion, that we must choose between the advantages and the good—we should thrust science away, disheartened; we should shut ourselves up in voluntary ignorance; above all, we should decline all participation in the affairs of our country, leaving to the men of another time the burden and the responsibility of a choice so difficult.

Chapter 15

RECIPROCITY AGAIN

The protectionists ask, "Are we sure that the foreigner will purchase as much from us, as he will sell to us? What reason have we to think that the English producer will come to us rather than to any other nation on the globe to look for the productions he may need; and for productions equivalent in value to his own exportations to this country?"

We are surprised that men who call themselves peculiarly *practical*, reason independent of all practice.

In practice, is there one exchange in a hundred, in a thousand, in ten thousand perhaps, where there is a direct barter of product for product? Since there has been money in the world, has any cultivator ever said, "I wish to buy shoes, hats, advice, instruction, from that shoemaker, hatter, lawyer, and professor only, who will purchase from me just wheat enough to make an equivalent value?"

And why should nations impose such a restraint upon themselves?

How is the matter managed?

Suppose a nation deprived of exterior relations. A man has produced wheat. He throws it into the widest national circulation he can find for it, and receives in exchange, what? Some dollars; that is to say bills, bonds, infinitely divisible, by means of which it becomes lawful for him to withdraw from national circulation, whenever he thinks it advisable, and by just agreement, such articles as he may need or wish. In fine, at the end of the operation he will have withdrawn from the mass the exact equivalent of what he threw into it, and in value his consumption will precisely equal his production.

If the foreign exchanges of that nation are free, it is no longer into *national,* but into *general* circulation that each one throws his products, and from which he draws his returns. He has not to inquire whether what he delivers up for general circulation is purchased by a fellow-countryman or a foreigner; whether the goods he receives came to him from a Frenchman or an Englishman; whether the objects for which, in accordance with his needs, he, in the end, exchanges his bills, are made on this or that side of the Atlantic or the St. Lawrence. With each individual there is always an exact balance between what he puts into and what he draws out of the grand common reservoir; and if that is true of each individual, it is true of the nation in the aggregate. The only difference between the two cases is, that in the latter, each one is in a more extended market for both his sales and his purchases, and has consequently more chances of doing well by both.

This objection is made: "If every one should agree that they would not withdraw from circulation any of the products of a specified individual, he in turn would sustain the misfortune of being able to draw nothing out. The same of a nation."

ANSWER.—If the nation cannot draw out of the mass, it will no longer contribute to it: it will work for itself. It will be compelled to that which you would impose on it in advance: that is to say, isolation.

And this will be the ideal of prohibitive government. Is it not amusing that you inflict upon it, at once and already, the misfortune of this system, in the fear that it runs the risk of getting there some day without you?

Chapter 16

OBSTRUCTED RIVERS PLEAD FOR THE PROHIBITIONISTS

Some years ago, when the Spanish Cortes were discussing a treaty with Portugal on improving the course of the river Douro, a deputy rose and said, "If the Douro is turned into a canal, transportation will be made at a much lower price. Portuguese cereals will sell cheaper in Castile, and will make a formidable opposition to our *national labor.* I oppose the project unless the ministers engage to raise the tariff in such a way as to restore the equilibrium." The assembly found the argument unanswerable.

Three months later the same question was submitted to the Senate of Portugal. A noble hidalgo said: "Mr. President, the project is absurd. You post guards, at great expense, on the banks of the Douro, in order to prevent the introduction of Castilian cereals into Portugal, while, at the same time, you would, also, at great expense, facilitate their introduction. This is an inconsistency with which I cannot identify myself. Let the Douro pass on to our sons as our fathers left it to us."

Now, when it is proposed to alter and confine the course of the Mississippi, we recall the arguments of the Iberian orators, and say to ourselves, if the member from St. Louis was as good an economist as those of Valencia, and the representatives from New Orleans as powerful logicians as those of Oporto, assuredly the Mississippi would be left

"To sleep amid its forests dank and lone," for to improve the navigation of the Mississippi will favor the introduction of New Orleans products to the injury of St. Louis, and an inundation of the products of St. Louis to the detriment of New Orleans.

Chapter 17

A NEGATIVE RAILROAD

We have said that when, unfortunately, we place ourselves at the point of view of the producer's interest, we cannot fail to clash with the general interest, because the producer, as such, demands only *efforts, wants, and obstacles.*

When the Atlantic and Great Western Railway is finished, the question will arise, "Should connection be broken at Pittsburg?" This the Pittsburgers will answer affirmatively, for a multitude of reasons, but for this among others; the railroad from New York to St. Louis ought to have an interruption at Pittsburg, in order that merchandise and travellers compelled to stop in the city may leave in it fees to the hackmen, pedlars, errand-boys, consignees, hotel-keepers, etc.

It is clear, that here again the interest of the agent of labor is placed before the interest of the consumer.

But if Pittsburg ought to profit by the interruption, and if the profit is conformable with public interest, Harrisburg, Dayton, Indianapolis, Columbus, much more all the intermediate points, ought to demand stoppages, and that in the general interest, in the widely extended interest of national labor, for the more they are multiplied, the more will consignments, commissions, transportations, be multiplied on all points of the line. With this system we arrive at a railroad of successive stoppages, to a *negative railroad.*

Whether the protectionists wish it or not, it is not the less certain that the principle of restriction is the same as the principle of gaps, the sacrifice of the consumers to the producer, of the end to the means.

Chapter 18

THERE ARE NO ABSOLUTE PRINCIPLES

We cannot be too much astonished at the facility with which men resign themselves to be ignorant of what is most important for them to know, and we may feel sure that they have decided to go to sleep in their ignorance when they have brought themselves to proclaim this axiom: There are no absolute principles.

Enter the Halls of Congress. The question under discussion is whether the law shall interdict or allow international exchanges.

Mr. C****** rises and says:

"If you tolerate these exchanges, the foreigner will inundate you with his products, the English with cotton and iron goods, the Nova-Scotian with coal, the Spaniard with wool, the Italian with silk, the Canadian with cattle, the Swede with iron, the Newfoundlander with salt-fish. Industrial pursuits will thus be destroyed."

Mr. G***** replies:

"If you prohibit these exchanges, the varied benefits which nature has lavished on different climates will be, to you, as though they were not. You will not participate in the mechanical skill of the English, nor in the riches of the Nova-Scotian mines, in the abundance of Canadian pasturage, in the cheapness of Spanish labor, in the fervor of the Italian climate; and you will be obliged to ask through a forced production that which you might by exchange have obtained through a readier production."

Assuredly, one of the senators deceives himself. But which? It is well worth while to ascertain; for we are not dealing with opinions only. You stand at the entrance of two roads; you must choose; one of them leads necessarily to *misery.*

To escape from this embarrassment it is said: There are no absolute principles.

This axiom, so much in vogue in our day, not only serves laziness, it is also in accord with ambition.

If the theory of prohibition should prevail, or again, if the doctrine of liberty should triumph, a very small amount of law would suffice for our economic code. In the first case it would stand—*All foreign exchange is forbidden;* in the second, *All exchange with abroad is free,* and many great personages would lose their importance.

But if exchange has not a nature proper to itself; if it is governed by no natural law; if it is capriciously useful or injurious; if it does not find its spring in the good it accomplishes, its limit when it ceases to do good; if its effects cannot be appreciated by those who execute them; in one word, if there are no absolute principles, we are compelled to measure, weigh, regulate transactions, to equalize the conditions of labor, to look for the level of profits—colossal task, well suited to give great entertainments, and high influence to those who undertake it.

Here in New York are a million of human beings who would all die within a few days, if the abundant provisioning of nature were not flowing towards this great metropolis.

Imagination takes fright in the effort to appreciate the immense multiplicity of articles which must cross the Bay, the Hudson, the Harlem, and the East rivers, to-morrow, if the lives of its inhabitants are not to become the prey of famine, riot, and pillage. Yet, as we write, all are sleeping; and their quiet slumbers are not disturbed for a moment by the thought of so frightful a perspective. On the other hand, forty-five States and Territories have worked to-day, without concert, without mutual understanding, to provision New York. How is it that every day brings in what is needed, neither more nor less, to this gigantic market? What is the intelligent and secret power which presides over the astonishing regularity of movements so complicated—a regularity in which each one has a faith so undoubting, though comfort and life are at stake.

This power is an *absolute principle,* the principle of freedom of operation, the principle of free conduct.

We have faith in that innate light which Providence has placed in the hearts of all men, to which he has confided the preservation and improvement of our race-*interest* (since we must call it by its name), which is so active, so vigilant, so provident, when its action is free. What would become of you, inhabitants of New York, if a Congressional majority should take a fancy to substitute for this power the combinations of their genius, however superior it may be supposed to be; if they imagined they could submit this prodigious mechanism to its supreme direction, unite all its resources in their own hands, and decide when, where, how, and on what conditions everything should be produced, transported, exchanged, and consumed? Ah! though there may be much suffering within your bounds, though misery, despair, and perhaps hungry exhaustion may cause more tears to flow than your ardent charity can dry, it is probable, it is certain, we dare to affirm, that the arbitrary intervention of government would multiply these sufferings infinitely, and would extend to you all, those evils which at present are confined to a small portion of your number.

We all have faith in this principle where our internal transactions are concerned; why should we not have faith in the same principle applied to our international operations, which are, assuredly, less numerous, less delicate, and less complicated. And if it is not necessary that the Mayor and Common Council of New York should regulate our industries, weigh our change, our profits, and our losses, occupy themselves with the regulation of prices, equalize the conditions of our labor in internal commerce—why is it necessary that the custom-house, proceeding on its fiscal mission, should pretend to exercise protective action upon our exterior commerce?

Chapter 19

NATIONAL INDEPENDENCE

Among the arguments which are considered of weight in favor of the restriction system, we must not forget that drawn from national independence.

"What shall we do in case of war," say they, "if we have placed ourselves at the mercy of Great Britain for iron and coal?"

English monopolists did not fail on their side to exclaim, when the corn-laws were repealed, "What will become of Great Britain in time of war if she depends on the United States for food?"

One thing they fail to observe: it is that this sort of dependence, which results from exchange, from commercial operations, is a *reciprocal* dependence. We cannot depend on the foreigner unless the foreigner depends on us. This is the very essence of *society*. We do not place ourselves in a state of independence by breaking natural relations, but in a state of isolation.

Remark also: we isolate ourselves in the anticipation of war; but the very act of isolation is the commencement of war. It renders it more easy, less burdensome, therefore less unpopular. Let nations become permanent recipient customers each of the other, let the interruption of their relations inflict upon them the double suffering of privation and surfeit, and they will no longer require the powerful navies which ruin them, the great armies which crush them; the peace of the world will no longer be compromised by the caprice of a Napoleon or of a Bismarck, and war will disappear through lack of aliment, resources, motive, pretext, and popular sympathy.

We know well that we shall be reproached (in the cant of the day) for proposing interest, vile and prosaic interest, as a foundation for the fraternity of nations. It would be preferred that it should have its foundation in charity, in love, even in self-renunciation, and that, demolishing the material comfort of man, it should have the merit of a generous sacrifice.

When shall we have done with such puerile talk? When shall we banish charlatanry from science? When shall we cease to manifest this disgusting contradiction between our writings and our conduct? We hoot at and spit upon *interest*, that is to say, the useful, the right (for to say that all nations are interested in a thing, is to say that that thing is good in itself), as if interest were not the necessary, eternal, indestructible instrument to which Providence has entrusted human perfectibility. Would not one suppose us all angels of disinterestedness? And is it supposed that the public does not see with disgust that this affected language blackens precisely those pages for which it is compelled to pay highest? Affectation is truly the malady of this age.

What! because comfort and peace are correlative things; because it has pleased God to establish this beautiful harmony in the moral world; you are not willing that we should admire and adore His providence, and accept with gratitude laws which make justice the condition of happiness. You wish peace only so far as it is destructive to comfort; and liberty burdens you because it imposes no sacrifices on you. If self-renunciation has so many claims for you, who prevents your carrying it into private life? Society will be grateful to you for it, for some one, at least, will receive the benefit of it; but to wish to impose it on humanity as a principle is the height of absurdity, for the abnegation of everything is the sacrifice of everything—it is evil set up in theory.

But, thank Heaven, men may write and read a great deal of such talk, without causing the world to refrain on that account from rendering obedience to its motive-power, which is, whether they will or no, *interest*. After all, it is singular enough to see sentiments of the most sublime abnegation invoked in favor of plunder itself. Just see to what this ostentatious disinterestedness tends. These men, so poetically delicate that they do not wish for peace itself, if it is founded on the base interest of men, put their hands in the pockets of others, and, above all, of the poor; for what section of the tariff protects the poor?

Well, gentlemen, dispose according to your own judgment of what belongs to yourselves, but allow us also to dispose of the fruit of the sweat of our brows, to avail ourselves of exchange at our own pleasure. Talk away about self-renunciation, for that is beautiful; but at the same time practice a little honesty.

Chapter 20

HUMAN LABOR—NATIONAL LABOR

To break machines, to reject foreign merchandise—are two acts proceeding from the same doctrine.

We see men who clap their hands when a great invention is made known to the world, who nevertheless adhere to the protective system. Such men are highly inconsistent.

With what do they upbraid freedom of commerce? With getting foreigners more skilful or better situated than ourselves to produce articles, which, but for them, we should produce ourselves. In one word, they accuse us of damaging national labor.

Might they not as well reproach machines for accomplishing, by natural agents, work which, without them, we could perform with our own arms, and, in consequence, damaging human labor?

The foreign workman who is more favorably situated than the American laborer, is, in respect to the latter, a veritable economic machine, which injures him by competition. In the same manner, a machine which executes a piece of work at a less price than can be done by a certain number of arms, is, relatively to those arms, a true competing foreigner, who paralyzes them by his rivalry.

If, then, it is needful to protect national labor against the competition of foreign labor, it is not less so, to protect human labor against the rivalry of mechanical labor.

So, he who adheres to the protective policy, if he has but a small amount of logic in his brain, must not stop when he has prohibited foreign products; he must farther proscribe the shuttle and the plough.

And that is the reason why we prefer the logic of those men who, declaiming against the invasion of exotic merchandise, have, at least, the courage to declaim as well against the excess of production due to the inventive power of the human mind.

Hear such a Conservative:—"One of the strongest arguments against liberty of commerce, and the too great employment of machines, is, that very many workmen are deprived of work, either by foreign competition, which is destructive to their manufactures, or by machines, which take the place of men in the workshops."

This gentleman perfectly sees the analogy, or rather, let us say, the identity, existing between importations and machines; that is the reason he proscribes both: and truly there is some pleasure in having to do with reasonings, which, even in error, pursue an argument to the end.

Let us look at the difficulty in the way of its soundness.

If it be true, *a priori,* that the domain of *invention* and that of labor cannot be extended, except at the expense of one or the other, it is in the place where there are most machines, Lancaster or Lowell, for example, that we shall meet with the fewest *workmen.* And if, on the contrary, we prove *a fact,* that mechanical and hand work co-exist in a greater degree among wealthy nations than among savages, we must necessarily conclude that these two powers do not exclude each other.

It is not easy to explain how a thinking being can taste repose in presence of this dilemma:

Either—"The inventions of man do not injure labor, as general facts attest, since there are more of both among the English and Americans than among the Hottentots and Cherokees. In that case I have made a false reckoning, though I know neither where nor when I got astray. I should commit the crime of treason to humanity if I should introduce my error into the legislation of my country."

Or else—"The discoveries of the mind limit the work of the arms, as some particular facts seem to indicate; for I see daily a machine do the labor of from twenty to a hundred workmen, and thus I am forced to prove a flagrant, eternal, incurable antithesis between the intellectual and physical ability of man; between his progress and his comfort; and I cannot forbear saying that the Creator of man ought to have given him either reason or arms, moral force, or brutal force, but that he has played with him in conferring upon him opposing faculties which destroy one another."

The difficulty is pressing. Do you know how they get rid of it? By this singular apothegm:

"In political economy there are no absolute principles."

In intelligible and vulgar language, that means: "I do not know where is the true nor the false; I am ignorant of what constitutes general good or evil; I give myself no trouble about it. The only law which I consent to recognize, is the immediate effect of each measure upon my personal comfort."

No absolute principles! You might as well say, there are no absolute facts; for principles are only the summing up of well proven facts.

Machines, importations, have certainly consequences. These consequences are good or bad. On this point there may be difference of opinion. But whichever of these we adopt, we express it in one of these two *principles:* "machines are a benefit," or "machines are an evil." "Importations are favorable," or "importations are injurious." But to say "there are no principles," is the lowest degree of abasement to which the human mind can descend; and we confess we blush for our country when we hear so monstrous a heresy uttered in the presence of the American people, with their consent; that is to say, in the presence and with the consent of the greater part of our fellow-citizens, in order to justify Congress for imposing laws on us, in perfect ignorance of the reasons for them or against them.

But then we shall be told, "destroy *the sophism;* prove that machines do not injure *human labor,* nor importations *national industry.*"

In an essay of this nature such demonstrations cannot be complete. Our aim is more to propose difficulties than to solve them; to excite reflection, than to satisfy it. No conviction of the mind is well acquired, excepting that which it gains by its own labor. We will try, nevertheless, to place it before you.

The opponents of importations and machines are mistaken, because they judge by immediate and transitory consequences, instead of looking at general and final ones.

The immediate effect of an ingenious machine is to economize, towards a given result, a certain amount of handwork. But its action does not stop there: inasmuch as this result is obtained with less effort, it is given to the public for a lower price; and the amount of the savings thus realized by all the purchasers, enables them to procure other gratifications—that is to say, to encourage handwork in general, equal in amount to that subtracted from the special handwork lately improved upon—so that the level of work has not fallen, though that of gratification has risen. Let us make this connection of consequences evident by an example.

Suppose that in the United States ten millions of hats are sold at five dollars each: this affords to the hatters' trade an income of fifty millions. A machine is invented which allows hats to be afforded at three dollars each. The receipts are reduced to thirty millions, admitting that the consumption does not increase. But, for all that, the other twenty millions are not subtracted from *human labor.* Economized by the purchasers of hats, they will serve them in satisfying other needs, and by consequence will, to that amount, remunerate collective industry. With these two dollars saved, John will purchase a pair of shoes, James a book, William a piece of furniture, etc. Human labor, in the general, will thus continue to be encouraged to the amount of fifty millions; but this sum, beside giving the same number of hats as before, will add the gratifications obtained by the twenty millions which the machine has spared. These gratifications are the net products which America has gained by the invention. It is a gratuitous gift, a tax, which the genius of man has imposed on Nature. We do not deny that, in the course of the change, a certain amount of labor may have been *displaced;* but we cannot agree that it has been destroyed, or even diminished. The same holds true of importations.

We will resume the hypothesis. America makes ten millions of hats, of which the price was five dollars each. The foreigner invaded our market in furnishing us with hats at three dollars. We say that national labor will be not at all diminished. For it will have to produce to the amount of thirty millions, in order to pay for ten millions of hats at three dollars. And then there will remain to each purchaser two dollars saved on each hat, or a total of twenty millions, which will compensate for other enjoyments; that is to say, for other work. So the total of labor remains what it was; and the supplementary enjoyments, represented by twenty millions economized on the hats, will form the net profit of the importations, or of free trade.

No one need attempt to horrify us by a picture of the sufferings, which, in this hypothesis, will accompany the displacement of labor. For if prohibition had never existed, labor would have classed itself in accordance with the law of exchange, and no displacement would have taken place. If, on the contrary, prohibition has brought in an artificial and unproductive kind of work, it is prohibition, and not free trade, which is responsible for the inevitable displacement, in the transition from wrong to right.

Unless, indeed, it should be contended that, because an abuse cannot be destroyed without hurting those who profit by it, its existence for a single moment is reason enough why it should endure forever.

Chapter 21

RAW MATERIAL

It is said that the most advantageous commerce consists in the exchange of manufactured goods for raw material, because this raw material is a spur to *national labor.*

And then the conclusion is drawn, that the best custom-house regulation would be that which should give the utmost possible facility to the entry of *raw material,* and oppose the greatest obstacles to articles which have received their first manipulation by labor.

No sophism of political economy is more widely spread than the foregoing. It supports not only the protectionists, but, much more, and above all, the pretended liberalists. This is to be regretted; for the worst which can happen to a good cause is not to be severely attacked, but to be badly defended.

Commercial freedom will probably have the fate of all freedom; it will not be introduced into our laws until after it has taken possession of our minds. But if it be true that a reform must be generally understood, in order that it may be solidly established, it follows that nothing can retard it so much as that which misleads public opinion; and what is more likely to mislead it than those writings which seem to favor freedom by upholding the doctrines of monopoly?

Several years ago, three large cities of France—Lyons, Bordeaux, and Havre—were greatly agitated against the restrictive policy. The nation, and indeed all Europe, was moved at seeing a banner raised, which they supposed to be that of free trade. Alas! it was still the banner of monopoly; of a monopoly a little more niggardly, and a great deal more absurd, than that which they appeared to wish to overturn. Owing to the sophism which we are about to unveil, the petitioners merely reproduced the doctrine of *protection to national labor,* adding to it, however, another folly.

What is, in effect, the prohibitive system? Let us listen to the protectionist: "Labor constitutes the wealth of a people, because it alone creates those material things which our necessities demand, and because general comfort depends upon these."

This is the principle.

"But this abundance must be the product of *national labor.* Should it be the product of foreign labor, national labor would stop at once."

This is the mistake. (See the close of the last chapter.)

"What shall be done, then, in an agricultural and manufacturing country?"

This is the question.

"Restrict its market to the products of its own soil, and its own industry."

This is the end proposed.

"And for this end, restrain by prohibitive duties the entrance of the products of the industry of other nations."

These are the means.

Let us reconcile with this system that of the petition from Bordeaux.

It divided merchandise into three classes:

"The first includes articles of food, and *raw material free from all human labor. A wise economy would require that this class should not be taxed.*"

Here there is no labor; consequently no protection.

"The second is composed of articles which have undergone *some preparation.* This preparation warrants us *in charging it with some tax.*"

Here protection commences, because, according to the petitioners, *national labor* commences.

"The third comprises perfected articles which can in no way serve national labor; we consider these the most taxable."

Here, labor, and with it protection, reach their maximum.

The petitioners assert that foreign labor injures national labor; this is *the error* of the prohibitive school.

They demanded that the French market should be restricted to French *labor;* this is the *end* of the prohibitive system.

They insisted that foreign labor should be subject to restriction and taxation; these are the *means* of the prohibitive system.

What difference, then, is it possible to discover between the petitioners of Bordeaux and the advocate of American restriction? One alone: the greater or less extent given to the word *labor.*

The protectionist extends it to everything—so he wishes to *protect* everything.

"Labor constitutes *all* the wealth of a people," says he; "to protect national industry, *all* national industry, manufacturing industry, *all* manufacturing industry, is the idea which should always be kept before the people." The petitioners saw no labor excepting that of manufacturers; so they would admit that alone to the favors of protection. They said:

"Raw material is *devoid of all human labor.* For that reason we should not tax it. Fabricated articles can no longer occupy national labor. We consider them the most taxable."

We are not inquiring whether protection to national labor is reasonable. The protectionist and the Bordelais agree upon this point, and we, as has been seen in the preceding chapters, differ from both.

The question is to ascertain which of the two—the protectionists or the raw-materialists of Bordeaux—give its just acceptation to the word "labor."

Now, upon this ground, it must be said, the protectionist is, by all odds, right; for observe the dialogue which might take place between them:

The Protectionist: "You agree that national labor ought to be protected. You agree that no foreign labor can be introduced into our market without destroying therein an equal amount of our national labor. Yet you assert that there is a host of merchandise possessed of *value* (since it sells), which is, however, free from *human labor.* And, among other things, you name wheat, corn, meats, cattle, lard, salt, iron, brass, lead, coal, wool, furs, seeds, etc. If you can prove to me that the value of these things is not due to labor, I will agree that it is useless to protect them. But, again, if I demonstrate to you that there is as much labor in a hundred dollars' worth of wool as in a hundred dollars' worth of cloth, you must acknowledge that protection is as much due to the one as to the other. Now, why is this bag of wool worth a hundred dollars? Is it not because that sum is the price of production? And is the price of production anything but that which it has been necessary to distribute in wages, salaries, manual labor, interest, to all the workmen and capitalists who have concurred in producing the article?"

The Protectionist: "Without doubt Nature *creates* the *elements* of all things; but it is labor which produces their *value*. I was wrong myself in saying that labor creates material objects, and this faulty phrase has led the way to many other errors. It does not belong to man, either manufacturer or cultivator, to *create*, to make something out of nothing; if, by *production*, we understand *creation*, all our labors will be unproductive; that of merchants more so than any other, except, perhaps, that of law-makers. The farmer has no claim to have *created* wheat, but he may claim to have created its *value:* he has transformed into wheat substances which in no wise resembled it, by his own labor with that of his ploughmen and reapers. What more does the miller effect who converts it into flour, the baker who turns it into bread? Because man must clothe himself in cloth, a host of operations is necessary. Before the intervention of any human labor, the true raw materials of this product (cloth) are air, water, gas, light, the chemical substances which must enter into its composition. These are truly the raw materials which are *untouched by human labor;* therefore, they are of no *value,* and I do not think of protecting them. But a first labor converts these substances into hay, straw, etc., a second into wool, a third into thread, a fourth into cloth, a fifth into clothing—who will dare to say that every step in this work is not *labor,* from the first stroke of the plough, which begins, to the last stroke of the needle, which terminates it? And because, in order to secure more celerity and perfection in the accomplishment of a definite work, such as a garment, the labors are divided among several classes of industry, you wish, by an arbitrary distinction, that the order of succession of these labors should be the only reason for their importance; so much so that the first shall not deserve even the name of labor, and that the last work pre-eminently, shall alone be worthy of the favors of protection!" The Raw-Materialist: "It is true, that in regard to wool, you may be right. But a bag of wheat, an ingot of iron, a quintal of coal—are they the produce of labor? Did not Nature create them?"

The Raw-Materialist: "Yes, we begin to see that wheat no more than wool is entirely devoid of human labor; but, at least, the agriculturist has not, like the manufacturer, done all by himself and his workmen; Nature aids him, and if there is labor, it is not all labor in the wheat."

The Protectionist: "But all its *value* is in the labor it has cost. I admit that Nature has assisted in the material formation of wheat. I admit even that it may be exclusively her work; but confess that I have controlled it by my labor; and when I sell you some wheat, observe this well: that it is not the work of *Nature* for which I make you pay, but *my own;* and, on your supposition, manufactured articles would be no more the product of labor than agricultural ones. Does not the manufacturer, too, rely upon Nature to second him? Does he not avail himself of the weight of the atmosphere in aid of the steam-engine, as I avail myself of its humidity in aid of the plough? Did he create the laws of gravitation, of correlation of forces, of affinities?"

The Raw-Materialist: "Come, let the wool go too. But coal is assuredly the work, and the exclusive work, of Nature, *unaided by any human labor.*"

The Protectionist: "Yes, Nature made coal, but *labor* makes its value. Coal had no *value* during the thousands of years during which it was hidden, unknown, a hundred feet below the soil. It was necessary to look for it there—that is a *labor:* it was necessary to transport it to market; that is another *labor:* and once more, the price which you pay for it in the market is nothing else than the remuneration for these labors of digging and transportation."

We see that thus far the protectionist has all the advantage on his side; that the value of raw material, as well as that of manufactured material, represents the expense of production, that is to say, of *labor;* that it is impossible to conceive of a material possessed of value while totally unindebted to human labor; that the distinction which the raw-materialists make is wholly futile, in theory; that, as a basis for an unequal division of *favors,* it would be iniquitous in practice; because the result would be that one-third of the people, engaged in manufactures, would obtain the sweets of monopoly, for the reason that they produced *by labor,* while the other two-thirds, that is to say the agriculturists, would be abandoned to competition, under pretext that they produced without labor.

It will be urged that it is of more advantage to a nation to import the materials called raw, whether they are or are not the product of labor, and to export manufactured articles.

This is a strongly accredited opinion.

"The more abundant raw materials are," said the petition from Bordeaux, "the more manufactories are multiplied and extended." It said again, that "raw material opens an unlimited field of labor to the inhabitants of the country from which it is imported."

"Raw material," said the other petition, that from Havre, "being the aliment of labor, must be submitted to a *different system,* and admitted at once at the lowest duty." The same petition would have the protection on manufactured articles reduced, not one after another, but at an undetermined time; not to the lowest duty, but to twenty per cent.

"Among other articles which necessity requires to be abundant and cheap," said the third petition, that from Lyons, "the manufacturers name all raw material."

This all rests on an illusion. We have seen that all *value* represents labor. Now, it is true that labor increases ten-fold, sometimes a hundred-fold, the value of a rough product, that is to say, expands ten-fold, a hundred-fold, the products of a nation. Thence it is reasoned, "The production of a bale of cotton causes workmen of all classes to earn one hundred dollars only. The conversion of this bale into lace collars raises their profits to ten thousand dollars; and will you dare to say that the nation is not more interested in encouraging labor worth ten thousand than that worth one hundred dollars?"

We forget that international exchanges, no more than individual exchanges, work by weight or measure. We do not exchange a bale of cotton for a bale of lace collars, nor a pound of wool in the grease for a pound of wool in cashmere; but a certain value of one of these things *for an equal value* of the other. Now to barter equal value against equal value is to barter equal work against equal work. It is not true, then, that the nation which gives for a hundred dollars cashmere or collars, gains more than the nation which delivers for a hundred dollars wool or cotton.

In a country where no law can be adopted, no impost established, without the consent of those whom this law is to govern, the public cannot be robbed without being first deceived. Our ignorance is the "raw material" of all extortion which is practised upon us, and we may be sure in advance that every sophism is the forerunner of a spoliation. Good public, when you see a sophism, clap your hand on your pocket; for that is certainly the point at which it aims. What was the secret thought which the shipowners of Bordeaux and of Havre, and the manufacturers of Lyons, conceived in this distinction between agricultural products and manufactured articles?

"It is principally in this first class (that which comprehends raw material *unmodified by human labor*)," said the Raw-Materialists of Bordeaux, "that the chief aliment of our merchant marine is found. At the outset, a wise economy would require that this class should not be taxed. The second (articles which have received some preparation) may be charged; the third (articles on which no more work has to be done) we consider the most taxable."

"Consider," said those of Havre, "that it is indispensable to reduce all raw materials one after another to the lowest rate, in order that industry may successively bring into operation the naval forces which will furnish to it its first and indispensable means of labor." The manufacturers could not in exchange of politeness be behind the ship-owners; so the petition from Lyons demanded the free introduction of raw material, "in order to prove," said they, "that the interests of manufacturing towns are not always opposed to those of maritime ones!"

True; but it must be said that both interests were, understood as the petitioners understood them, terribly opposed to the interests of the country, of agriculture, and of consumers.

See, then, where you would come out! See the end of these subtle economical distinctions! You would legislate against allowing *perfected* produce to traverse the ocean, in order that the much more expensive transportation of rough materials, dirty, loaded with waste matter, may offer more employment to our merchant service, and put our naval force into wider operation. This is what these petitioners termed *a wise economy.* Why did they not demand that the firs of Russia should be brought to them with their branches, bark, and roots; the gold of California in its mineral state, and the hides from Buenos Ayres still attached to the bones of the tainted skeleton?

Industry, the navy, labor, have for their end, the general good, the public good. To create a useless industry, in order to favor superfluous transportation; to feed superfluous labor, not for the good of the public, but for the expense of the public—this is to realize a veritable begging the question. Work, in itself, is not a desirable thing; its result is; all work without result is a loss. To pay sailors for carrying useless waste matter across the sea is like paying them for skipping stones across the surface of the water. So we arrive at this result: that all economical sophisms, despite their infinite variety, have this in common, that they confound the means with the end, and develop one at the expense of the other.

Chapter 22

METAPHORS

Sometimes a sophism dilates itself, and penetrates through the whole extent of a long and heavy theory. More frequently it is compressed, contracted, becomes a principle, and is completely covered by a word. A good man once said: "God protect us from the devil and from metaphors!" In truth, it would be difficult to say which of the two creates the more evil upon our planet. It is the demon, say you; he alone, so long as we live, puts the spirit of spoliation in our hearts. Yes; but he does not prevent the repression of abuses by the resistance of those who suffer from them. *Sophistry* paralyzes this resistance. The sword which malice puts in the assailant's hand would be powerless, if sophistry did not break the shield upon the arm of the assailed; and it is with good reason that Malebranche has inscribed at the opening of his book, "Error is the cause of human misery."

See how it comes to pass. Ambitious hypocrites will have some sinister purpose; for example, sowing national hatred in the public mind. This fatal germ may develop, lead to general conflagration, arrest civilization, pour out torrents of blood, draw upon the land the most terrible of scourges—*invasion*. In every case of indulgence in such sentiments of hatred they lower us in the opinion of nations, and compel those Americans, who have retained some love of justice, to blush for their country. Certainly these are great evils; and in order that the public should protect itself from the guidance of those who would lead it into such risks, it is only necessary to give it a clear view of them. How do they succeed in veiling it from them? It is by *metaphor.* They alter, they force, they deprave the meaning of three or four words, and all is done.

Such a word is *invasion* itself. An owner of an American furnace says, "Preserve us from the *invasion* of English iron." An English landlord exclaims, "Let us repel the *invasion* of American wheat!" And so they propose to erect barriers between the two nations. Barriers constitute isolation, isolation leads to hatred, hatred to war, and war to *invasion*. "Suppose it does," say the two sophists; "is it not better to expose ourselves to the chance of an eventual *invasion,* than to accept a certain one?" And the people still believe, and the barriers still remain.

Yet what analogy is there between an exchange and an *invasion?* What resemblance can possibly be established between a vessel of war, which comes to pour fire, shot, and devastation into our cities, and a merchant ship, which comes to offer to barter with us freely, voluntarily, commodity for commodity?

As much may be said of the word *inundation.* This word is generally taken in bad part, because *inundations* often ravage fields and crops. If, however, they deposit upon the soil a greater value than that which they take from it; as is the case in the inundations of the Nile, we might bless and deify them as the Egyptians do. Well! before declaiming against the inundation of foreign produces, before opposing to them restraining and costly obstacles, let us inquire if they are the inundations which ravage or those which fertilize? What should we think of Mehemet Ali, if, instead of building, at great expense, dams across the Nile for the purpose of extending its field of inundation, he should expend his money in digging for it a deeper bed, so that Egypt should not be defiled by this *foreign* slime, brought down from the Mountains of the Moon? We exhibit precisely the same amount of reason, when we wish, by the expenditure of millions, to preserve our country—From what? The advantages with which Nature has endowed other climates.

Among the metaphors which conceal an injurious theory, none is more common than that embodied in the words *tribute, tributary.*

These words are so much used that they have become synonymous with the words *purchase, purchaser,* and one is used indifferently for the other.

Yet a *tribute* or *tax* differs as much from *purchase* as a theft from an exchange, and we should like quite as well to hear it said, "Dick Turpin has broken open my safe, and has *purchased* out of it a thousand dollars," as we do to have it remarked by our sage representatives, "We have paid to England the *tribute* for a thousand gross of knives which she has sold to us."

For the reason why Turpin's act is not a *purchase* is, that he has not paid into my safe, with my consent, value equivalent to what he has taken from it, and the reason why the payment of five hundred thousand dollars, which we have made to England, is not a *tribute,* is simply because she has not received them gratuitously, but in exchange for the delivery to us of a thousand gross of knives, which we ourselves have judged worth five hundred thousand dollars.

But is it necessary to take up seriously such abuses of language? Why not, when they are seriously paraded in newspapers and in books?

Do not imagine that they escape from writers who are ignorant of their language; for one who abstains from them, we could point you to ten who employ them, and they persons of consideration—that is to say, men whose words are laws, and whose most shocking sophisms serve as the basis of administration for the country.

A celebrated modern philosopher has added to the categories of Aristotle, the sophism which consists in including in one word the begging of the question. He cites several examples. He should have added the word *tributary* to his vocabulary. In effect the question is, are purchases made abroad useful or injurious? "They are injurious," you say. And why? "Because they make us *tributary* to the foreigner." Here is certainly a word which presents as a fact that which is a question.

How is this abusive trope introduced into the rhetoric of monopolists?

Some specie *goes out of a country* to satisfy the rapacity of a victorious enemy—other specie, also, goes out of a country to settle an account for merchandise. The analogy between the two cases is established, by taking account of the one point in which they resemble one another, and leaving out of view that in which they differ.

This circumstance, however,—that is to say, non-reimbursement in the one case, and reimbursement freely agreed upon in the other—establishes such a difference between them, that it is not possible to class them under the same title. To deliver a hundred dollars *by compulsion* to him who says "Stand and deliver," or *voluntarily* to pay the same sum to him who sells you the object of your wishes—truly, these are things which cannot be made to assimilate. As well might you say, it is a matter of indifference whether you throw bread into the river or eat it, because in either case it is bread *destroyed.* The fault of this reasoning, as in that which the word *trib-ute* is made to imply, consists in founding an exact similitude between two cases on their points of resemblance, and omitting those of differ-ence.

Chapter 23

CONCLUSION

All the sophisms we have hitherto combated are connected with one single question: the restrictive system; and, out of pity for the reader, we pass by acquired rights, untimeliness, misuse of the currency, etc., etc.

But social economy is not confined to this narrow circle. Fourierism, Saint-Simonism, communism, mysticism, sentimentalism, false philanthropy, affected aspirations to equality and chimerical fraternity, questions relative to luxury, to salaries, to machines, to the pretended tyranny of capital, to distant territorial acquisitions, to outlets, to conquests, to population, to association, to emigration, to imposts, to loans, have encumbered the field of science with a host of parasitical *sophisms,* which demand the hoe and the sickle of the diligent economist. It is not because we do not recognize the fault of this plan, or rather of this absence of plan. To attack, one by one, so many incoherent sophisms which sometimes clash, although more frequently one runs into the other, is to condemn one's self to a disorderly, capricious struggle, and to expose one's self to perpetual repetitions.

How much we should prefer to say simply how things are, without occupying ourselves with the thousand aspects in which the ignorant see them! To explain the laws under which societies prosper or decay, is virtually to destroy all sophistry at once. When La Place had described all that can, as yet, be known of the movements of the heavenly bodies, he had dispersed, without even naming them, all the astrological dreams of the Egyptians, Greeks, and Hindoos, much more surely than he could have done by directly refuting them through innumerable volumes. Truth is one; the book which exposes it is an imposing and durable monument:

Il brave les tyrans avides,
Plus hardi que les Pyramides
Et plus durable que l'airain.

Error is manifold, and of ephemeral duration; the work which combats it does not carry within itself a principle of greatness or of endurance.

But if the power, and perhaps the opportunity, have failed us for proceeding in the manner of La Place and of Say, we cannot refuse to believe that the form which we have adopted has, also, its modest utility. It appears to us especially well suited to the wants of the age, to the hurried moments which it can consecrate to study.

A treatise has, doubtless, an incontestable superiority; but upon condition that it be read, meditated upon, searched into. It addresses itself to a select public only. Its mission is, at first, to fix, and afterwards to enlarge, the circle of acquired knowledge.

The refutation of vulgar prejudices could not carry with it this high bearing. It aspires only to disencumber the route before the march of truth, to prepare the mind, to reform public opinion, to blunt dangerous tools in improper hands. It is in social economy above all, that these hand-to-hand struggles, these constantly recurring combats with popular errors, have a true practical utility.

But there are sciences which exercise upon the public an influence proportionate with the light of the public itself, not from knowledge accumulated in a few exceptional heads, but from that which is diffused through the general understanding. Such are morals, hygiene, social economy, and in countries which men belong to themselves, politics. It is of these sciences, above all, that Bentham might have said: "That which spreads them is worth more than that which advances them." Of what consequence is it that a great man, a God even, should have promulgated moral laws, so long as men, imbued with false notions, take virtues for vices, and vices for virtues? Of what value is it that Smith, Say, and, according to Chamans, economists of all schools, have proclaimed the superiority of liberty to restraint in commercial transactions, if those who make the laws and those for whom the laws are made, are convinced to the contrary. We might arrange the sciences under two classes. The one, strictly, can be known to philosophers only. They are those whose application demands a special occupation. The public profit by their labor, despite their ignorance of them. They do not enjoy the use of a watch the less, because they do not understand mechanics and astronomy. They are not the less carried along by the locomotive and the steamboat through their faith in the engineer and the pilot. We walk according to the laws of equilibrium without being acquainted with them.

These sciences, which are well named social, have this peculiarity: that for the very reason that they are of a general application, no one confesses himself ignorant of them. Do we wish to decide a question in chemistry or geometry? No one pretends to have the knowledge instinctively; we are not ashamed to consult Draper; we make no difficulty about referring to Euclid.

But in social science authority is but little recognized. As such a one has to do daily with morals, good or bad, with hygiene, with economy, with politics reasonable or absurd, each one considers himself skilled to comment, discuss, decide, and dogmatize in these matters.

Are you ill? There is no good nurse who does not tell you, at the first moment, the cause and cure of your malady.

"They are humors," affirms she; "you must be purged."

But what are humors? and are these humors?

She does not trouble herself about that. I involuntarily think of this good nurse when I hear all social evils explained by these common phrases: "It is the superabundance of products, the tyranny of capital, industrial plethora," and other idle stories of which we cannot even say: *verba et voces praetereaque nihil:* for they are also fatal mistakes.

From what precedes, two things result—

1st. That the social sciences must abound in sophistry much more than the other sciences, because in them each one consults his own judgment or instinct alone.

2d. That in these sciences sophistry is especially injurious, because it misleads public opinion where opinion is a power—that is, law.

Two sorts of books, then, are required by these sciences; those which expound them, and those which propagate them; those which show the truth, and those which combat error.

It appears to us that the inherent defect in the form of this little Essay—*repetition*—is that which constitutes its principal value.

In the question we have treated, each sophism has, doubtless, its own set form, and its own range, but all have one common root, which is, *"forgetfulness of the interests of man, insomuch as they forget the interests of consumers."* To show that the thousand roads of error conduct to this generating sophism, is to teach the public to recognize it, to appreciate it—to distrust it under all circumstances.

After all, we do not aspire to arouse convictions, but doubts.

We have no expectation that in laying down the book, the reader shall exclaim: *"I know."* Please Heaven he may be induced to say, *"I am ignorant."*

"I am ignorant, for I begin to believe there is something delusive in the sweets of Scarcity."

"I am no longer so much edified by the charms of Obstruction."

"Effort without Result no longer seems to me so desirable as Result without Effort."

"It may probably be true that the secret of commerce does not consist, as that of arms does, *in giving and not receiving,* according to the definition which the duellist in the play gives of it."

"I consider an article is increased in value by passing through several processes of manufacture; but, in exchange, do two equal values cease to be equal because the one comes from the plough and the other from the power-loom?"

"I confess that I begin to think it singular that humanity should be ameliorated by shackles, or enriched by taxes: and, frankly, I should be relieved of a heavy weight, I should experience a pure joy, if I could see demonstrated, which the author assures us of, that there is no incompatibility between comfort and justice, between peace and liberty, between the extension of labor and the progress of intelligence."

"So, without feeling satisfied by his arguments, to which I do not know whether to give the name of reasoning or of objections, I will interrogate the masters of the science."

Let us terminate by a last and important observation this monograph of sophisms. The world does not know, as it ought, the influence which sophistry exerts upon it. If we must say what we think, when the Right of the Strongest was dethroned, sophistry placed the empire in the Right of the Most Cunning; and it would be difficult to say which of these two ty-rants has been the more fatal to humanity.

Men have an immoderate love for pleasure, influence, position, power—in one word, for wealth.

And at the same time men are impelled by a powerful impulse to pro-cure these things at the expense of another. But this other, which is the public, has an inclination not less strong to keep what it has acquired, provided it can and knows how. Spoliation, which plays so large a part in the affairs of the world, has, then, two agents only: Strength and Cun-ning; and two limits: Courage and Right.

Power applied to spoliation forms the groundwork of human savagism. To retrace its history would be to reproduce almost entire the history of all nations—Assyrians, Babylonians, Medes, Persians, Egyptians, Greeks, Romans, Goths, Franks, Huns, Turks, Arabs, Moguls, Tartars—without counting that of the Spaniards in America, the English in India, the French in Africa, the Russians in Asia, etc., etc.

But, at least, among civilized nations, the men who produce wealth have become sufficiently numerous and sufficiently strong to defend it.

Is that to say that they are no longer despoiled? By no means; they are robbed as much as ever, and, what is more, they despoil one another. The agent alone is changed; it is no longer by violence, but by stratagem, that the public wealth is seized upon.

In order to rob the public, it must be deceived. To deceive it, is to per-suade it that it is robbed for its own advantage; it is to make it accept ficti-tious services, and often worse, in exchange for its property. Hence soph-istry, economical sophistry, political sophistry, and financial sophistry—and, since force is held in check, sophistry is not only an evil, it is the par-ent of other evils. So it becomes necessary to hold it in check, *in its turn,* and for this purpose to render the public more acute than the cunning; just as it has become more peaceful than the strong.

www.ingramcontent.com/pod-product-compliance
Lightning Source LLC
LaVergne TN
LVHW092030060326
832903LV00058B/493